T0092262

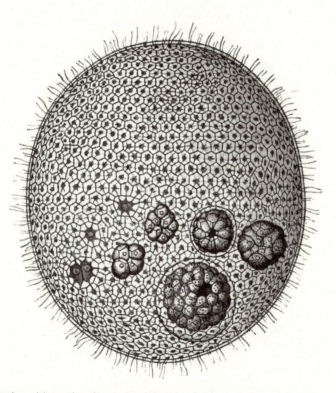

Volvox globator Ehrenberg. An adult asexual colony, highly magnified. The hexagonal areas represent the gelatinous coats of the individual cells in surface view. The thin common envelope of the whole colony is seen round the circumference. In the hinder half of the colony are seen two of the large asexual reproductive cells, and various stages of their development into daughter-colonies. The two most advanced daughter-colonies have already secreted a common envelope of their own. (After A. Lang.)

THE INDIVIDUAL IN THE ANIMAL KINGDOM

JULIAN S. HUXLEY
FOREWORD BY RICHARD GAWNE AND
JACOBUS J. BOOMSMA

The MIT Press
Cambridge, Massachusetts
London, England

This book was set in Bembo Book MT Pro by New Best-set Typesetters Ltd. Printed and bound in the United States of America.

Library of Congress Cataloging-in-Publication Data

Names: Huxley, Julian, 1887–1975, author.
Title: The individual in the animal kingdom / Julian S. Huxley ; foreword by Richard Gawne and Jacobus J. Boomsma.
Description: Cambridge, Massachusetts : The MIT Press, [2022] | Originally published in Cambridge, Eng., by Cambridge University Press and New York by G. P. Putnam's Sons, 1912. | Includes bibliographical references and index.
Identifiers: LCCN 2021031216 | ISBN 9780262045377 (hardcover)
Subjects: LCSH: Life (Biology)—Philosophy. | Individuality.
Classification: LCC QH501 .H88 2022 | DDC 113/.8—dc23
LC record available at https://lccn.loc.gov/2021031216

10   9   8   7   6   5   4   3   2   1

CONTENTS

LIST OF ILLUSTRATIONS

Volvox globator Ehrenberg   *Frontispiece*

Figs. 2, 10, and 11 are reproduced from the *Encyclopaedia Britannica* (eleventh edition); figs. 6 and 12 are from Lankester's *Treatise on Zoology*, Vol. 11, and Scott's *Structural Botany* respectively, by kind permission of Messrs A. & C. Black; fig. 7 is from West's *British Freshwater Algae* (Camb. Univ. Press); and fig. 13 is from Weismann's *Evolution Theory*, by permission of Mr Edward Arnold.

FOREWORD

RICHARD GAWNE AND JACOBUS J. BOOMSMA

I. INTRODUCTION

Julian Huxley's *The Individual in the Animal Kingdom* is a book about the major transitions in evolution (MTEs), *avant la lettre*. When it was published in 1912, the modern evolutionary synthesis had not occurred, the central dogma of molecular biology was unknown, and the experimental methods we now take for granted were impossible to fathom. Biology has clearly come a long way since the early twentieth century, so some might infer they will find little of value in this monograph. The standard response would be that standing on the shoulders of past giants is the only way to see into the distant scientific horizon, but Newton's aphorism only holds true if the footings we are perched upon are sound. Sifting through the historical literature, it becomes apparent that many works are, in fact, best forgotten. However, intellectual archaeology can sometimes help to uncover long-lost gems with unique vantage points that modern biologists can use to advance their research. In our view, *The Individual in the Animal Kingdom* is one of these treasures.

A close reading of enduring works from the past often confirms that their insights have been incorporated into the contemporary literature in a refined and expanded form. This is cumulative scientific progress at its finest. Revisiting

The Individual in the Animal Kingdom, one gets the impression that research on the MTEs has charted a less desirable course. Huxley's hypotheses are so strikingly clear and original that our present-day understanding of the MTEs appears to represent modest, incremental progress. Indeed, one could argue that the field has actually regressed in some respects due to a neglect of rigorous, falsifiable theorizing and subsequent empirical testing. The current state of affairs is partly attributable to the lack of attention *The Individual in the Animal Kingdom* has received. Our goal in overseeing this reissue is to give this groundbreaking work a second hearing. Despite the fact that it is now over one hundred years old, and despite Huxley himself being just twenty-five years old when it was pressed (figure F.1), the book remains highly relevant to contemporary research on the MTEs. The primary aim of our introduction is to demonstrate this point in order to encourage biologists to read the body of the text in its original form. An exhaustive historical contextualization of the book that purports to be free from evaluative judgments will have to wait for another day.

Current thinking[1,2] is that the MTEs—understood as a series of events where formerly independent entities joined forces to form a more organizationally complex whole—were first described by John Maynard Smith and Eörs Szathmáry in their 1995 book,[3] *The Major Transitions in Evolution*, and a *Nature* paper published in the same year.[4] In a semantic sense, this is correct because the *term* "major evolutionary transition" had not previously been used in this specific context. However, the *fact* that life has increased in organizational complexity over time—the central tenet of the modern MTE framework—had been appreciated for well over a century. More specifically, many of the ideas currently circulating in the literature were

Figure F.1
Photograph courtesy of Woodson Research Center, Fondren Library,
Rice University. Signature provided by and used with the permission
of Cold Spring Harbor Laboratory Archives, NY.

first incisively formulated by Huxley in *The Individual in the Animal Kingdom*. Yet, it is important to emphasize that even in 1912, the basic idea that organisms have evolved different levels of hierarchical complexity was not new.

For example, by the 1800s biologists were regularly highlighting parallels between multicellular "cell-states" and human societies. Much of this work was focused on how cellular order is maintained physiologically,[5,6] but more than a few researchers of this period began to explore the dangerous possibility that such systems might provide a desirable model of government for human societies. As the better-known controversies between Herbert Spencer and T. H. Huxley illustrate,[7] evolutionary biology in the late nineteenth century was highly political, and some of the mistaken ideas developed during this time eventually helped to spawn disastrous totalitarian regimes in the twentieth century. Thankfully, the scientific discourse on cells and societies eventually became more rigorous following the formalization of cell theory. During the early 1900s, researchers began to realize that the biological part–whole relationship between metazoan cells and the integrated organisms they form is not comparable to the sociopolitical dynamics that exist between free citizens and their states.[8] That being said, important parallels between "insect states" and cell states do exist, and Huxley was among the first to see these.

By noting the affinities between metazoan bodies and caste-differentiated insect colonies, Huxley was following in the footsteps of the myrmecologist William Morton Wheeler. One year before *The Individual in the Animal Kingdom* appeared, Wheeler hypothesized that the same type of common descent relationship that holds between differentiated metazoan cells and protists exists between the members of an ant colony and their solitary sister clades.[9] Elaborating on Darwin's concept of family selection as an explanation of insect neuters, he concluded that ant colonies are *organisms*, meaning their lineages have transitioned to a higher type of organization that is more hierarchically complex than obligate multicellularity. Wheeler's original "superorganism" was a carefully crafted theoretical concept that used ontogenetic considerations to identify lineages that have transitioned to a new level of social organization. Despite its sophistication, his concept failed to garner significant interest, and was eventually misappropriated when ecologists incorrectly claimed that communities and ecosystems undergo development, and subsequently exhibit organism-like harmony.[10] However, related topics such as levels-of-organization and the interactions between the component parts and larger wholes of organisms were regularly explored throughout much of the twentieth century. A substantial portion of this work was aimed at understanding causation in animal development from a physiological perspective, but later studies added an explicit evolutionary component as well.[11]

None of this historical work, including Huxley's book, was cited by Maynard Smith and Szathmáry, giving many the incorrect impression that their conceptual framework had no precedent. Their dialectic stood in contrast to Leo Buss's book *The Evolution of Individuality*,[12] which was more explicitly informed by natural history, had a wider range of taxonomic coverage, and relied on a broader, albeit sometimes eclectic, range of primary sources. Although he was more thorough in his details, Buss also failed to acknowledge Huxley's pioneering work. Part of the reason why the phenomena we now refer to as the MTEs have been "discovered" multiple times is that subsequent generations used different terms to describe these events—including "individuality," "organismality," "organizational complexity," and most recently, "major transitions" facilitated by changes in encoded genetic information—without making an effort to determine how these ideas relate to one other. Everyone seems to have assumed that they were first to coherently describe the evolution of organizational complexity, and since none of these research programs used Huxley's logic as a foundational stepping-stone, the literature has fragmented into parallel traditions that provide varying levels of insight into different features of the MTEs.

The Individual in the Animal Kingdom is written in a narrative style that was common in the early 1900s, when biology was essentially a qualitative discipline. Because of this, some of Huxley's language will be unfamiliar to modern readers (see glossary), and the text may seem almost poetic at times because it lacks the rigid structural format we have become accustomed to in recent decades. Nevertheless, in terms of the actual scientific content, one is hard-pressed to identify claims that are false, and this is what truly matters. It would be absurd to ignore the works of Shakespeare or Chaucer because they deviate from the rules of the *Chicago Manual of Style*, but we often appear more than willing to do something similar when an older work of science does not strictly conform to modern semantic conventions. There is no doubt that Huxley's book takes more effort to read and digest than a contemporary journal article that consists of boilerplate declarative sentences interspersed with summary figures and uninterpreted p-values. However, we believe that biologists—along with a much wider circle of academics—who take the time to read the rather modest number of pages that comprise *The Individual in the Animal Kingdom* will find it to be a rewarding experience.

Throughout his book, Huxley strives for theoretical precision and conceptual coherence. Paragraph by paragraph, page by page, he is pulling his readers toward a very specific set of conclusions. Huxley begins by carefully defining what he means by "individuality" and then applies the concept to a wide array of terrestrial and aquatic organisms that span multiple kingdoms. For each of the taxa examined, he reviews the basic biology of example species and then indicates what level of organizational complexity the lineages have achieved. Huxley was a skilled naturalist who was not content with producing a mere descriptive taxonomy of individuality. From the start, he argues that transitions in hierarchical organization must have been *adaptive*, and develops testable inferences about the ultimate and proximate causes that facilitated these events. Huxley's overarching system of concepts, and the wide-ranging logical principles that he applied to all known levels of hierarchical complexity were so far ahead of their time that they continue to stand tall in the company of modern work on the MTEs.

In contrast to *The Individual in the Animal Kingdom*, Maynard Smith and Szathmáry's book offers almost no hypotheses about the necessary conditions that precede the origin of an MTE. The lack of emphasis on ultimate selective forces and their enabling proximate mechanisms that began with their descriptive framework continues to characterize a substantial portion of current MTE research. Instead of formulating falsifiable hypotheses that seek to explain *why* transitions in organizational complexity take place, many biologists remain content to note *that* they have occurred. To make matters worse, the term "major transition" is now used in an array of contexts that have little or nothing to do with hierarchical changes in organization. Once the MTE-concept *sensu* Maynard Smith and Szathmáry had acquired a certain degree of professional cachet, researchers began using it to describe almost every imaginable type of change that was deemed to be evolutionarily important, or simply worthy of study. The precedent for this promiscuous usage was arguably set by Maynard Smith and Szathmáry themselves. As others have correctly pointed out,[13] the criteria they used to identify the MTEs were vaguely defined, and as a result the events they considered to be major transitions were a highly heterogenous menagerie from the outset.

One of the refreshing features of Huxley's approach is his implicit emphasis of the fact that MTEs promote what we would now call adaptive radiations. Conceiving things in this way enabled him to carve the natural world along joints that had long been obvious to naturalists. It appears that Huxley's primary goals were to: (i) formulate a generalized predictive theory of progress in organizational complexity, and (ii) demonstrate that natural selection has produced both quantitative and discontinuous MTE-like changes via common principles that apply across *all* levels of hierarchical organization. As Niels Bohr demonstrated with his correspondence principle, the best scientific theories tend to integrate phenomena that occur at different scales in nature. Just as Bohr was able to show that quantum discontinuities do not contradict continuum mechanics in the macroscopic world, Huxley managed to reconcile the gradual evolutionary change produced by natural selection with the seemingly conflicting fact that MTEs constitute dichotomous breaks in organizational complexity. We know of no published studies that undermine his highly parsimonious logic. If such rebuttals truly do not exist, Huxley's 1912 book deserves to be ranked among the most important works in organismal evolutionary theory published after the *Origin of Species*.

II. INDIVIDUALITY AND ORGANISMALITY

Huxley's definition of individuality explicitly considers part–whole relationships and is linked to, but not synonymous with, the complementary concept of organismality. An entity can be regarded as an individual when observation reveals that it is not a component of a larger integrated whole with its own unique ontogeny. Parts have no ability to survive or reproduce on their own, and therefore are not individuals in and of themselves. Their existence is dependent upon the context of the more inclusive unit they belong to, and this is what prevents them from qualifying as biological individuals. All organisms are individuals, but organisms always come in grades (a discrete hierarchical measure), while individuality can also vary in degree (a continuous intralevel measure) within these grades. To give an example, an isolated cardiac cell is not an individual because it is unable to naturally persist when isolated from the rest of a body—it is merely one part of an organismal whole that belongs to a higher (multicellular) grade. Yet, other single-celled entities such as *Paramecium* protists can readily survive and reproduce on their own. They are not functional parts of anything, but autonomous, free-living organisms, albeit of a lower-level (unicellular) grade. And, autonomous *Paramecium* ciliates exhibit a higher degree of biological individuality than other protists which lack some of their more derived features.

Huxley eventually sets himself to the task of making his abstract definitions useful to working biologists, and comprehensible to the general public. He notes that "living matter always tends to group itself into . . . closed, independent systems with harmonious parts" (p. xlvi), and adds that this tells us something important about the boundaries that demarcate the various types of organizational complexity. More specifically, he proceeds by suggesting that an entity can be regarded as an individual if it: (1) is a closed system, meaning it has a gatekeeping mechanism that differentiates self (the organism) from non-self (the abiotic and biotic environment); (2) exhibits some degree of independence from these environments; and (3) can be observed to possess differentiated internal parts, which (4) interact harmoniously with one another to promote the survival of a larger whole that is capable of reproduction. All organisms exhibit these characteristic features of individuality and, at first glance, they appear to be expressed in a completely continuous manner, such that some might be more independent from their environment, or show more internal division of labor, than others. However, Huxley recognized that "even the most perfect quantitative gradation from one condition to another is no guarantee that the two conditions shall not be qualitatively different" (p. 58). In other words, not all variation in individuality is continuous. Organisms vary in the hierarchical level at which they are closed, which translates into discrete differences in the "grade of individuality" or level of organismal complexity they have attained. While the continuous degrees of individuality Huxley describes address progression and secondary reduction of morphological complexity, behavioral repertoires, etc., the discontinuous differences in grade he identifies will be our primary focus because they are *evolutionarily irreversible*, and have a direct connection to the modern MTE framework.

The idea that organisms—be it cells, multicellular individuals, or organismal colonies—are closed, hierarchically arranged systems is integral to Huxley's analysis. At any given level, closure requires the evolution of physical structures, or other mechanisms that perform isolating functions. Cell membranes are a canonical anatomical feature that ensures closure, and higher-level analogues include animal epithelia, and colonial gestalts. Once closure has been established, a new boundary between self and nonself is created, and what was formerly nothing more than lower-level units staying together can become a unified organism (pp. 6–8; 10–11) that has gained a novel and higher level of independence from its environment. Closure is also necessary for the evolution of an internal division of labor, understood here as functional specialization of subordinate units that needs to take place in order for grade transitions to occur. This appears to be driven by an "all in the same boat" type of principle, where cooperation is the only viable option, and functional complementarity (think rowing and bailing out water) is obviously beneficial to all. Once some members of the newly closed unit begin to acquire specialized functional roles, the logical implication is that the fitness of the higher-level individual should increase.

Huxley's theory of how the MTEs originated is parsimonious insofar as it assumes that, provided closure is secured, complementary internal specialization has the potential to arise. This is subsequently expected to promote the evolution of analogous forms of higher-level adaptive organization across all grades. The idea that MTEs have to be irreversible is another striking feature of *The Individual in the Animal Kingdom*. At several points in the text, Huxley implies that transitional events are akin to what we might now call a "one-way street," and crucially, he attributes the progressive quality of the MTEs to the fact that natural selection must sometimes have favored organisms that increased their independence from the environment (compared to their recent ancestors) by becoming closed at higher organizational levels.

Once a lineage evolves a new closure mechanism and transitions in grade, the parts of the larger organismic whole are no longer able to survive and reproduce independently, as their free-living ancestors once did. Huxley illustrates this point with a memorable witticism, noting that "if you divide a man into two by cutting off his hand, the working of . . . [the man] . . . is rendered less effective," while the functioning of the hand is "stopped for ever" (p. 8). In contrast, he clearly recognizes that the part–whole relationship is one of

convenience, not necessity, when groups are merely same-
generation aggregates. The parts of aggregates keep their
"perfect independence as they would have if living solitary"
(p. 75), making them different from organismal structures
bound by clonal or family ties that keep the whole together
across subsequent cohorts of cells or individuals. Huxley also
points out that the macroscopic mutualisms known at the
time are always aggregative, which preliminarily suggests that
their capacity to transition in organismal grade is severely constrained (pp. 95–96).

III. THE MAJOR TRANSITIONS IN EVOLUTION

From the opening pages of *The Individual in the Animal Kingdom*, it is evident that young Julian Huxley was an extraordinarily well-rounded academic who took inspiration from
many sources, including the philosophers Friedrich Nietzsche
and Henri Bergson. Perhaps most notably, he extracted the
idea of "closed systems" from Bergson's literary prose, and
transformed it into a robust scientific concept (pp. 6–8). Like
other scholars, particularly in his generation, Huxley inevitably found various works of art, philosophy, and literature
enjoyable, insightful, or even inspiring. However, *The Individual in the Animal Kingdom* is very much the work of an
expert biologist that is firmly rooted in the empiricist tradition of comparative zoology. It would be absurd for scientists
to dismiss Huxley's book as speculative storytelling because
it contains quotes from *Through the Looking Glass* (p. 65), and
similarly, there is nothing inherently problematic about using
the abstract thoughts of philosophers as a springboard for
formulating deeper insights into the biological world. Being
widely read is a virtue, not a vice.

Following Darwin and Weismann, Huxley emphasized that if we want to understand gradual changes and discontinuous transitions in organizational complexity (figure F.2), we should avoid thinking in terms of a Great Chain of Being that treats human individuality as the pinnacle that all other lineages must be measured against. He stated without qualification that "any view of animal individuality as a whole . . . must not take man and mammals as the single starting-point whence we could logically work backwards to the rest of the organic world." Instead, he emphasized that we should treat the organizational complexity of our bodies and those of our close relatives as "an ending instead of a beginning" that is merely "one . . . among many" (p. 25).

Ultimately, one could argue it was Huxley's recognition that humans are not special biologically and his knowledge of comparative natural history that allowed him to formulate the first coherent MTE framework. His upbringing in a scientifically minded family, and training in neo-Darwinian comparative zoology at Oxford likely enabled him to see that increases in organizational complexity involving abrupt, discontinuous breaks with ancestral body plans and life histories evolved by the same principles across distantly related lineages. This emphasis on common causes appears to have helped Huxley avoid the anthropomorphic conclusion that metazoan multicellularity must necessarily be the evolutionary pinnacle of fully integrated hierarchical complexity. That is to say, he kept his mind open to the possibility that nonhuman taxa such as caste-differentiated social insects exhibit a higher level of organismal organization than primate bodies and the much looser societies they can form. However, he was keenly aware that such transitions always come at the expense of the liberty of the "persons" involved (pp. 90–91, 104, 108),

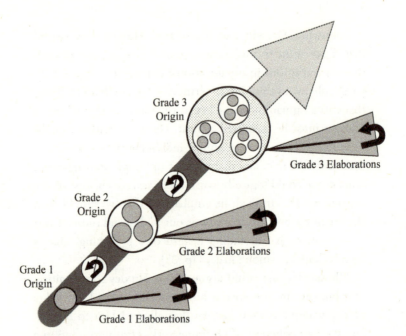

Figure F.2
Transitions in hierarchical complexity between grades are
evolutionarily irreversible, but elaborations within a grade are not.
A recurring theme of *The Individual in the Animal Kingdom* is that the
MTEs have a hierarchically progressive quality, meaning that once a
lineage has crossed into a new type of organization, its component
parts can no longer survive on their own. Conversely, within any given
level of organization, organisms often form facultative aggregates
with other entities. In these assemblages, internal division of labor
is usually weak or non-existent compared to multicellular bodies or
caste-differentiated colonies. In other words, aggregates are often
reversible elaborations because most individuals that belong to such
collectives do not specialize to become parts. Among strictly solitary
organisms, elaborations and regressions in organizational complexity
are both possible. The latter are almost always a consequence of
parasitic or mutualistic overspecialization (p. 100). In the figure, the
trajectory along the large arrow represents irreversible transitions in
organizational grade. Within each grade, lateral increases and possible
decreases in complexity are illustrated by slightly tilted horizontal
triangles.

implying that they should be avoided in the context of human
government (p. 116).

Recent commentators on the MTEs have tended to shun the
subject of evolutionary progress almost entirely, with some
going so far as to suggest that organizational complexity cannot be discretely ranked.[14,15] However, Huxley was adamant
that there are fundamentally distinct grades of organization
in the natural world, which have evolved in a stepwise fashion
from lower to higher via natural selection. His idea of progress
having been both qualitative and quantitative is demonstrated
by his assertion that, "in spite of many side-ventures," organisms can be "arranged in a single main series in which certain
characters are manifested more clearly and more thoroughly at
the top than at the bottom" (see figure F.2). Throughout the
text, he compiles evidence—surprisingly solid even by current standards—which indicates that there is a directionality
or arrow of organismal complexity. None of his arguments
rely on untestable vitalistic or orthogenetic claims about evolutionary forces trying to achieve some sort of preordained
goal state. They are based on the empirical observation that
life was originally simple and increased in hierarchical complexity over time. Grade transitions, then, represent occasional events where natural selection rewards new forms of
closure that facilitates higher-level division of labor and promotes increased environmental independence. In Huxley's
terms, these improvements enhance an individual's "survival
value," relative to its peer individuals.

Huxley identified three discrete grades of organization in
the natural world: (1) unicellular protists, (2) multicellular
organisms, and (3) (super)organismal colonies. The hierarchical nestedness of these grades is obvious because all grade 2
individuals are evolutionarily derived collectives of grade 1

organisms, and all grade 3 individuals are assemblies comprised of grade 2 multicellular parts. Huxley's taxonomy differs from the modern MTE framework in that it includes all unicellular organisms in the first grade. We now know that protists arose from a specific endosymbiotic fusion of prokaryote cells, and accordingly, there is unanimous agreement that the origin of the eukaryotic cell constituted a mutualistic MTE from simpler cellular ancestors. Given the technological limitations of his era, Huxley cannot be faulted for lumping all single-celled organisms into a common grade. However, his general reasoning about the evolution of hierarchical complexity is coherent enough that *The Individual in the Animal Kingdom* can easily be brought up to date by adding a prokaryote grade 0 category to his system that would set the Bacteria and Archaea apart (figure F.2) from the Eukaryota. Had Huxley known about the endosymbiotic origin of the last eukaryote common ancestor, he likely would have regarded it as a rare and welcome complement to his inference that macroscopic ectosymbioses do not induce grade-level transitions (pp. 95–96).

THE FIRST ORGANIZATIONAL GRADE

By the time *The Individual in the Animal Kingdom* was published, the basic principles of cell theory were widely accepted. Among other things, there was broad agreement that the cell is the basic compositional unit of life, and that all and only living entities are made-up of cells.[16] Huxley elaborated on established cell theory by formulating hypotheses about the origin of life (pp. 20, 24), and examining how the cellular composition of organisms has facilitated or constrained their evolution by natural selection. He argues that the formation of the first cells was a necessary precondition for life on Earth, meaning that the advent of the cell *caused* the origin

of life as we know it. A century later, our increased knowledge of viruses has complicated the situation slightly, but
many would still consider the first cell closure as the start of
independently organized life. Although Huxley was unable
to directly test his hypotheses about how life began, he used
existing knowledge to make a number of plausible inferences
about how the cell might have originated. It is worth empha-
sizing that these claims were empirically falsifiable, because
vitalist theories of life were still being promoted in the early
twentieth century. Huxley's tepid praise of Bergson could
give the impression that he might have been sympathetic to
such ideas, but there is no textual evidence for this in his 1912
book. As described above, he simply formalized Bergson's idea
of biological closure, and in his hands, this has no vitalistic
baggage.

According to Huxley, the early chemical constituents of
life had physical properties that predisposed them to form
closed cell-like structures (p. 38). These primitive cells were
the first organized lifeforms, which natural selection acted on
to produce more hierarchically complex, but equally closed,
multicellular organisms. As Huxley put it:

> [The cell] was like Benjamin Franklin's kite, bringing lightning
> down from heaven, but it did more than that, for it provided a
> permanent resting-place on earth where individuality could stay,
> could gather strength and develop upwards (p. 49).

The first cells were—and all extant unicellular prokaryotes
and eukaryotes continue to be—among the most successful
organisms on the planet. This is partly because the basic architecture of the cell can be modified through the inclusion of
structures that allow specialized functions to be performed.
Huxley used the following metaphor to illustrate this point:

Life finds in the cell the ground-plan for her first mansion—a one-roomed hut. You may change your one-roomed plan from round to square, from square to oblong, and you will not have improved it: but add a chimney and windows, and at once, though still but one room, it is something better (p. 46).

Although unicellular organisms have evolved beating cilia, photoreceptors, pseudopodia, and countless other structures that increase their fitness under specific conditions, the grade 1 body plan also has significant limitations. Almost all single-celled organisms are small, which makes them subject to a set of unique microscale challenges that haphazardly push and pull them in random directions (pp. 66–67).[17] Some unicellular organisms have evolved methods of mitigating the effects of Brownian dynamics, but these forces can never be truly defeated at smaller size ranges. Huxley therefore concludes that single-celled lifeforms are independent from their environment in only the barest possible sense. As he writes:

[I]t is impossible to think of any single-celled animal swimming against the most sluggish river as it is to imagine a butterfly poised steady in a twenty-knot gale (p. 67).

The fact that unicellular organisms have relatively little agency relegates them to the lowest grade of individuality and limits the number of niches they can adapt to inhabit. Given these constraints, occasional tendencies towards multicellularity could obviously have been promoted by natural selection under the right environmental conditions. When selective forces transformed some protist lineages into obligately multicellular organisms, a fundamentally new type of body plan that Huxley refers to as the second grade of individuality independently arose (figure F.2) several times.

THE SECOND ORGANIZATIONAL GRADE

As Huxley described it, grade 2 transitions occur when cells begin "joining up . . . together so that each preserved a considerable measure of independence, and was yet subordinated to the good of the whole." The end result is a "type of structure, [where] the individual is built up out of a number of cells instead of one" (p. 68). Huxley's grade 2 individuals are strictly synonymous with what we would now refer to as obligately multicellular organisms. However, his views about the evolution of multicellularity did not set an immediate precedent for clear thinking that was passed down without interruption from 1912. Particularly in recent decades, researchers have consistently misunderstood multicellularity as a seamless gradient that includes both facultative aggregates of unicellular organisms, and the obligate forms of multicellularity observed in metazoans, land plants, some fungi, and some algae. The idea that there are important, empirically verifiable differences between obligate and facultative multicellularity, and between aggregative and adhesive multicellularity, used to be canonical.[18] However, these fundamental distinctions have only recently resurfaced[19]—an interesting case of a field returning to its roots after embracing alternative notions that proved unproductive.

Although multicellularity can be adaptive, the fact that so many protist species currently exist suggests that transitions to grade 2 must have been difficult, and hence rarely occurred. This being so, one could wonder why unicellular organisms have not employed other strategies to inhabit new niches, such as increasing their size by multiple orders of magnitude, while remaining within the first grade. Huxley realized that size increases of this type are generally impossible because the skewed surface area to volume ratio of excessively large cells precludes provisioning via diffusion. Once more, he employs a compelling metaphor to make his point:

> If the English Nation, with population advancing by leaps . . . were not able to build harbours and provide dock labourers as quick as she bred man, all the wheat in Canada . . . would not keep her from starvation for the simple reason that it could not get in. . . . [F]or each four-fold increase of transport workers there is a sixteen-fold increase within of the mouths to be fed (pp. 41–42).

Given the problems with surface area to volume scaling mentioned in this passage, it should come as no surprise that small clumps of cells sometimes emerged, and have been rewarded by natural selection. Huxley summarized the steps involved in this process as follows:

> Suppose that instead of separating from each other after each division, the cells remain connected. The result will be a colony of cells each one like all its fellows. If division of labour sets in later among the cells, they are rendered mutually dependent, and the colony is transformed into a true individual, which is obviously a higher order than the cell. It has attained what may be termed the second grade of individuality (pp. 46–47).

Two parts of this passage are of particular interest. First, Huxley appears to indicate that irreversible transitions to multicellularity are explicitly facilitated by adhesive properties of clonal cells, a contention that took a century to be formally confirmed by comparative data.[20,21] Second, he emphasizes that the advent of division of labor necessarily increased mutual dependence. Indeed, only in organisms that exhibit grade 2 organization are the component parts so interdependent that somatic cells are unable to escape the body and survive independently as solitary protists. As Huxley notes, reductions in

complexity due to overspecialization in parasitic or mutualistic niches may occur, but these have never returned a multicellular eukaryote to a protist.

Contemporary research has tended to leave the irrevers-
ibility of MTE events unexplored, partly due to a general
reluctance to see any sort of progressiveness in evolution, even
when it comes to something as straightforward as the compositional structure of organisms. Claims such as "Multicellular
organisms are composed of *simpler* cellular parts" or the assertion that "Metazoans are *more complex* than choanoflagellates"
are scientific statements about hierarchical composition, not
value judgments. As John Tyler Bonner described the situation, nearly everyone agrees that the first organisms were
unicellular, and species such as our own are either the most
complex, or somewhere near to this. Yet, "it is considered
bad form to take this as any kind of progression," presumably
because there is some sort of "subconscious desire among us
to be democratic even about our position in the great scale of
being."[22] Huxley lacked the blind spot identified by Bonner
and this significantly enhanced his theorizing. He recognized
that we often have legitimate reasons to speak of higher and
lower or simple and complex in biological contexts, and he
had no qualms acknowledging that some types of evolutionary change can be considered lasting advances in organizational integration. Clearly, however, such rankings are only
unambiguously justified when making comparisons between
phylogenetic crown groups that completed an MTE, and their
sister lineages that did not.

THE THIRD ORGANIZATIONAL GRADE

The gradual emergence of classification systems that confused
mere multicellular aggregations with true grade 2 organisms

seems to have been part of a wider trend that accelerated with
the sociobiology movement.[23] Continuing a theme that had
been prominent in the Chicago School of community ecology
for decades, behavioral and evolutionary researchers began to
downplay the importance of obligate co-dependence in the
classification of clones and families by advocating the use
of all-inclusive group definitions. This effectively obscured
the progressive features of the irreversible MTEs Huxley
identified. Continuum-style thinking had become deeply
entrenched by the time Maynard Smith and Szathmáry published their book,[24] and as a consequence few protested when
they lumped facultatively and obligately cooperative phenomena into categories such as "multicellular" and "society," as
if these terms were well-defined taxonomical designations.[25]
Retrospectively, it is apparent that the categorical boundaries
between social grades that Huxley recognized—which had
been generally accepted throughout most of the twentieth
century—were erased without any justification. This is not
a matter of semantics, or a trivial definitional quibble. The
vaguely defined sociobiological concepts Maynard Smith and
Szathmáry adopted have obscured the fundamental differences
between true, irreversible MTEs and facultative aggregations
of grade 1 or grade 2 organisms that do not constitute a lasting
advance in organizational complexity.

Groups of cells or multicellular individuals that come
together to form aggregates are normally able to leave and live
a solitary existence. Obligately multicellular organisms with
an internal germ–soma division of labor cannot do this—
their cells are forced to stay together because they are completely dependent upon the larger bodily collective. As both
Wheeler[26] and Huxley realized, there is a close parallel between
the developmental history of permanently multicellular

organisms and colonial social insects with permanently differentiated castes. Like metazoan bodies, these colonies are
tightly integrated units—albeit of a higher grade—and this
mutual dependence makes it impossible for individual animals
to function independently of the collective. Wilson's[27] continuum concept of eusociality masked this deep germline–soma
analogy, and encouraged the formation of misleading apples-
to-oranges comparisons because it did not recognize the fundamental distinctions between obligate social systems with a
higher-level ontogeny, and facultative aggregates that do not
develop in any meaningful sense of the term. In the end, Maynard Smith and Szathmáry's adoption of all-inclusive sociobiological gradient thinking made their account of the core
features of grade 3 organisms less incisive than Huxley's.

Thankfully, there is a way to escape this confusion. The
signal in the data quickly reemerges when we classify cellular
collectives such as *Dictyostelium* fruiting bodies with mere temporary division of labor as aggregates of grade 1 organisms.
The same transparency also re-emerges when we follow the
lead of Huxley and Wheeler when dealing with the functional
classification of the social insects.[28] To grasp how insect social
systems differ from one another, and to understand the ways
in which they are analogous to multicellular organisms, it is
necessary to: (i) abandon the broad-brush continuum categories, and give serious consideration to the importance of (ii)
closure and (iii) permanent, non-facultative division of labor.

Huxley followed Wheeler in acknowledging that ant colonies are grade 3 individuals and thus superorganismal (p. 108).
In their comparative thinking, grade 3 organisms first formed
closed colonies of core families, and then started to evolve
increasing specialization of their component individuals, leading to the same mutual codependence that is the hallmark of

germline and soma cells in grade 2 organisms. The key issue is that irreversible MTEs emerge when all colony members commit to either the germline or soma for life. Huxley drove this point home forcefully when he asked, "What is the meaning of a lonely bee and its actions when it comes back to find its hive destroyed?" (p. 7). Ultimately, the entity that replicates in grade 3 individuals is a specific colony lifecycle with a predictable higher-level ontogeny and replacement procedures of individual members, just like a solitary grade 2 organism reproduces a bodily lifecycle with a predictable development and turnover of somatic cells. At grade 3 MTE origins, closure at colony founding via a mated queen was a necessary condition of colony-level reproduction, just like zygote closure is for somatic body reproduction. However, in addition, natural selection must also forge an irreversible dichotomous break with the ancestral mode of reproducing by transforming formerly independent cell copies or offspring into soma (see "hierarchy" in the glossary).

In addition to recognizing that several insect lineages evolved organismal colonies, Huxley was aware that certain pelagic siphonophores have also transitioned to grade 3 organization, but was quick to add that other modular lineages of marine invertebrates have not done so, despite having some division of labor. The fact that he was able to generalize across taxa is impressive, and what made this possible was his commitment to a simple pair of theoretical principles: (i) MTEs only originated in ancestors with permanently closed social systems, and (ii) internal division of labor must be irreversibly advanced in order to qualify as a diagnostic MTE feature. Writing on the second idea, Huxley was clearly aware that division of labor in superorganisms can involve both germline (queen–worker) segregation and intra-somatic (e.g., worker–soldier) differentiation. He also grasped that the latter type of internal specialization was only known from ants and termites, and—in modular form—from cnidarians such as Portuguese man o' war.

Huxley's description of the basic biology of siphonophores is worth quoting at some length because it highlights how fundamental the concept of universal mutual dependence (or mutual parasitism for reciprocal efficiency benefits, as he writes on p. 103) was to his thinking about the MTEs between grades of individuality:

> In the majority of the Siphonophora, the persons of the colony have mostly only a historical individuality: some of them are sometimes so much modified and reduced that it has baffled all the zoologists to decide whether they are homologous with individuals or with mere appendages of individuals: and in function each is devoted so little to itself, so wholly to serving some particular need of the whole, that if one were separated from the rest, it would appear a perfectly useless and meaningless body to an investigator who did not know the whole to which it belonged. (p. 91)

The obligate, complex division of labor observed in siphonophores is particularly interesting because it illustrates that grade 3 transitions can occur in lineages that lack a complex central nervous system. At several points in *The Individual in the Animal Kingdom*, Huxley emphasizes that human brains are special, but advanced cognitive capacities are clearly neither necessary, nor sufficient for transitioning to the third organizational grade. This is vividly illustrated by the observation that brainless cnidarians have become grade 3 organisms, while our own societies remain, biologically speaking, within the second grade (pp. 108–109). In Appendix A (pp. 121–122), however, Huxley appears to add a puzzling wrinkle to the story by suggesting that human societies and some mutualisms *have* transitioned to the third organizational grade, albeit with a few caveats. The very brief tabulated entries appear to contradict claims made earlier in the text, where he argues in more detail that these systems have made only minor progress in this direction. Regardless, the crucial takeaway remains that Huxley was clearly aware that most offspring in ant and honeybee societies are developmentally somaticised into colony-parts, and that this system of biological organization would not be desirable for human social groups (p. 116).

IV. HISTORICAL REFLECTIONS AND FUTURE DIRECTIONS

Looking back at the scientific literature from the twentieth century, there are multiple examples of contributions that failed to receive adequate attention because they appeared in obscure journals or were the work of lesser-known practitioners. This is unfortunate, but also understandable given the sheer volume of research that was produced. In contrast, *The Individual in the Animal Kingdom*'s anonymity among contemporary biologists makes almost no sense at all. How did we manage to lose track of a Cambridge University Press monograph dedicated to a widely studied subject that was composed by one of the most revered scientists of the twentieth century? As we have shown, Huxley's book was not forgotten because it is filled with empirically inaccurate claims or dated theories that are no longer tenable in light of modern scientific knowledge. Although *The Individual in the Animal Kingdom*'s rapid descent into obscurity remains puzzling, three factors may help to explain how it slipped through the cracks.

First, most biologists are focused on emerging experimental methods and specific model organisms, which means that they often pay less attention to the work produced by those who came before them than physicists and chemists do. The fact that this trend has accelerated in recent decades is one of the most regrettable aspects of present-day biological research. Although modern technologies such as high-throughput genomic sequencing and their associated analysis pipelines have led to countless advances, it is important to remember that these are merely *tools*. Theories are the true bedrock of science, and in biology, the theory of evolution by natural selection continues to reign supreme. Synthetic theories and concepts function as a cross-generational matrix that keeps each science together over time. They are able to perform this function because they can remain valid, and hence relevant, no matter whether they were tested using techniques from the 1970s, 1990s, or 2010s. When our methods become the message, and the theoretical foundations of the field are ignored, the work of previous academic generations is inevitably lost, meaning we have no chance to build on their work or learn from their mistakes. In this light, the lack of attention *The Individual in the Animal Kingdom* has garnered over the years has almost certainly impacted the historical development of organismal biology. In particular, one could ask whether Wilson's *Sociobiology* and Maynard Smith and Szathmáry's *Major Transitions in Evolution* would have been written differently if earlier researchers had systematically tested and elaborated upon Huxley's framework.

Second, the trajectory of Huxley's career also seems to have played a role in the demise of his first book. Following the publication of *The Individual in the Animal Kingdom*, he embarked on a restless path through life, and was seldom settled academically. After a few years at Rice University, the University of Oxford, and King's College London, he effectively left academia in 1927 to focus on public dissemination

and synthesis of existing knowledge. Yet, he never lost his
passion for practical science and natural history, and remained
a figure of towering influence, particularly in Oxford, UK,
where his students and mentees obtained professorial positions.[29] Many believe that Huxley's crowning achievement
was his well-known book *Evolution: The Modern Synthesis*,[30]
but he also produced papers and monographs devoted to field
observations of birds and invertebrates, experimental work on
amphibian metamorphosis, and general essays on evolutionary theory and organismal scaling relationships, to name just
a few. Interestingly, Huxley seldom cited *The Individual in the
Animal Kingdom* in these works, even when a perfectly rea-
sonable opportunity to do so presented itself. Given his later
interest in *Drosophila* genetics—he dedicated his 1942 book to
Thomas Hunt Morgan, calling him a "many-sided leader in
biology's advance"[31]—it appears that he may have eventually
come to regard *The Individual in the Animal Kingdom* as a product of youthful fancy that was somehow incompatible with
newly arrived fashions, including the purportedly harder
reductionist biology of the day.

Third, science is in an almost constant state of flux, but the
rate and kind of innovation varies from decade to decade. In
biology, the first half of the twentieth century was a particularly
innovative period. Molecular biology, developmental genetics,
and the modern evolutionary synthesis all began during this
time, and these advances played a pivotal role in getting the
field to where it stands today. However, in retrospect it is
clear that the zeitgeist of the mid-twentieth century also led to
the abandonment of perfectly good practices that were previously the stock and trade of the profession. In developmental
biology, the entire tradition of comparative embryology was
jettisoned almost overnight, and although it was more erratic

and gradual, "old-fashioned" comparative natural history also declined. These approaches are prominently featured in *The Individual in the Animal Kingdom*, so many of Huxley's successors might have simply moved on to newer methods that appeared to promise more rapid career advancement.

One hundred ten years later, there is little question that, despite their obvious importance, the breakthroughs of the early twentieth century did not eliminate the need for careful observations, comparative phylogenetic inferences, and clear conceptual thinking within a broad frame of reference. Indeed, one could argue that these are some of biology's most essential methods that are always needed, and will never be replaced. Aside from the many empirical insights it contains, reading Huxley's 1912 book is worthwhile because it reveals that modern biological advances can sometimes lead the field down blind conceptual alleys. This is seldom apparent if we are only considering our immediate professional surroundings, or the recent history of a field. To understand whether a current direction of research is desirably incisive, it is often necessary to seek input from unbiased outside observers. No one is better suited to this task than our own academic forebears. However, when their wisdom is passed down through third-party summaries, the coherent insights they left behind may become distorted by misunderstandings, academic prejudice, and other forms of memetic drift. For this reason, it is important that we continue to read books such as *The Individual in the Animal Kingdom* in their original form. The giants of biology's past are still here to help shape the future of the field, but this can only happen if we pay the courtesy of reading their work.

★

We thank Thomas Pradeu and James DiFrisco for helpful comments on earlier drafts of this foreword, and Gerd Müller for his encouragement when this project was still in its infancy. Staff members at the American Philosophical Society, Cold Spring Harbor Laboratory, and Rice University graciously provided access to archival material. We also add special thanks to the production team at the MIT Press for making this reissue possible.

NOTES

1. Eörs Szathmáry, "Toward Major Evolutionary Transitions Theory 2.0," *Proceedings of the National Academy of Sciences* 112, no. 33 (2015): 10104–10111.

2. Stuart A. West, Roberta M. Fisher, Andy Gardner, and E. Toby Kiers. "Major Evolutionary Transitions in Individuality," *Proceedings of the National Academy of Sciences* 112, no. 33 (2015): 10112–10119.

3. John Maynard Smith and Eors Szathmáry, *The Major Transitions in Evolution* (Oxford: Oxford University Press, 1995).

4. Eörs Szathmáry and John Maynard Smith, "The Major Evolutionary Transitions," *Nature* 374, no. 6519 (1995): 227–232.

5. Thomas Pradeu, "Organisms or Biological Individuals? Combining Physiological and Evolutionary Individuality," *Biology & Philosophy* 31, no. 6 (2016): 797–817.

6. Andrew Reynolds, "The Theory of the Cell State and the Question of Cell Autonomy in Nineteenth and Early Twentieth-Century Biology," *Science in Context* 20, no. 1 (2007): 7.

7. James Elwick, "Herbert Spencer and the Disunity of the Social Organism," *History of Science* 41, no. 1 (2003): 35–72.

8. Pradeu, "Organisms or Biological Individuals?"

9. William Morton Wheeler, "The Ant-Colony as an Organism," *Journal of Morphology* 22, no. 2 (1911): 307–325.

10. Jacobus J. Boomsma and Richard Gawne, "Superorganismality and Caste Differentiation as Points of No Return: How the Major Evolutionary Transitions Were Lost in Translation," *Biological Reviews* 93, no. 1 (2018): 28–54.

11. John Tyler Bonner, *Cells and Societies* (Princeton, NJ: Princeton University Press, 1955).

12. Leo W. Buss, *The Evolution of Individuality* (Princeton, NJ: Princeton University Press, 1987).

13. Daniel W. McShea and Carl Simpson, "The Miscellaneous Transitions in Evolution," in *The Major Transitions in Evolution Revisited*, ed. Brett Calcott and Kim Sterelny, 19–34 (Cambridge, MA: MIT Press, 2011).

14. Paul W. Sherman, Eileen A. Lacey, Hudson K. Reeve, and Laurent Keller, "The Eusociality Continuum," *Behavioral Ecology* 6, no. 1 (1995): 102–108.

15. Peter Godfrey-Smith, *Darwinian Populations and Natural Selection* (Oxford: Oxford University Press, 2009).

16. Daniel J. Nicholson, "Biological Atomism and Cell Theory," *Studies in History and Philosophy of Science Part C: Studies in History and Philosophy of Biological and Biomedical Sciences* 41, no. 3 (2010): 202–211.

17. D. J. Nicholson, "On Being the Right Size, Revisited: The Problem with Engineering Metaphors in Molecular Biology," in *Philosophical Perspectives on the Engineering Approach in Biology: Living Machines?*, ed. S. Holm and M. Serban, 40–68. (London: Routledge, 2020).

18. Bonner, *Cells and Societies*.

19. Roberta M. Fisher, Charlie K. Cornwallis, and Stuart A. West, "Group Formation, Relatedness, and the Evolution of Multicellularity," *Current Biology* 23, no. 12 (2013): 1120–1125.

20. Fisher, Cornwallis, and West, "Group Formation, Relatedness, and the Evolution of Multicellularity."

21. William C. Ratcliff, R. Ford Denison, Mark Borrello, and Michael Travisano, "Experimental Evolution of Multicellularity," *Proceedings of the National Academy of Sciences* 109, no. 5 (2012): 1595–1600.

22. John Tyler Bonner, *The Evolution of Complexity by Means of Natural Selection* (Princeton, NJ: Princeton University Press, 1988), 5.

23. Edward O. Wilson, *Sociobiology: The New Synthesis* (Cambridge, MA: Harvard University Press, 1975).

24. Sherman et al., "The Eusociality Continuum."

25. Boomsma and Gawne, "Superorganismality and Caste Differentiation as Points of No Return."

26. Wheeler, "The Ant-Colony as an Organism."

27. Wilson, *Sociobiology*.

28. Boomsma and Gawne, "Superorganismality and Caste Differentiation as Points of No Return."

29. W. D. Provine, "England," in *The Evolutionary Synthesis*, ed. E. Mayr and W. B. Provine, 329–333 (Cambridge, MA: Harvard University Press, 1998).

30. Julian Huxley, *Evolution: The Modern Synthesis* (London: George Allen & Unwin,1942).

31. Huxley, *Evolution*.

GLOSSARY

RICHARD GAWNE AND JACOBUS J. BOOMSMA

Aggregation: May refer both to assemblies that "come together" and to assemblies that "stay together" due to internal division of labor, but only the obligation to stay together is a diagnostic feature of a grade transition, or MTE. The compositional units of aggregations that come together remain largely or completely homogenous, and as Huxley put it "nothing homogeneous can be an individual" (p. 7). The implication is that coming together enhances individuality rather little, and thus precludes the possibility of evolving into organismal wholes of higher-level unity (pp. 7–8).

Bacteria: Technological limitations prevented Huxley from differentiating between prokaryotic and eukaryotic unicellular organisms. Nevertheless, he was aware that bacteria have short generation times (pp. 25–26), are able to fix nitrogen (p. 129n5), possess "granules" that perform hereditary and assimilatory functions in the absence of a nucleus (p. 45), are asexual (p. 54), decay organic matter (p. 95), and likely bear a strong resemblance to the very first cells (pp. 43–44, 129).

Closure: A process whereby a boundary is created between organism and environment. Closure promotes environmental independence and enables natural selection to create an internal division of labor within organisms (p. 39). Huxley borrowed this concept from the philosopher Bergson, who infamously advocated the unscientific notion of *élan vital*. Accordingly, some have incorrectly accused Huxley of being a crypto-vitalist, even though he explicitly states that Bergson's vitalism is controversial (pp. xlv, 45) and can promote naïve inferences (p. 128n7). In Huxley's hands, closure is fully consistent with neo-Darwinism and the scientific method in general.

Colony: An aggregate of cells or multicellular units without internal division of labor, not a true organism. Forming a colony is necessary for achieving a higher level of individuality, but it is not sufficient because irreversible transitions require an internal germ–soma division of labor to evolve (pp. 75–76, 81, 89).

Community: A colony where cells or multicellular organisms stay together and have evolved irreversible division of labor. "Community" and "family" are synonyms as they were for Darwin. The parts of a community work for the whole and cannot survive independently (p. 28). Examples include ant colonies (p. 108) and cell colonies such as *Volvox*, which exhibit both germ–soma and within-soma differentiation (p. 80). This term is unrelated to ecological communities. Huxley realized that ecosystem-level communities are not closed, and thus lack individuality in any meaningful sense (p. 95). Note the difference between historical and modern uses of the terms "colony" (no division of labor implied then, division of labor assumed now) and "community" (obligate division of labor as part of secondarily extended family ontogeny implied then, now used only for ecological assemblies that have succession of participating species and individuals, but lack any form of organismal ontogeny).

Hierarchy: Grades of individuality are hierarchical in that grade 3 individuals (e.g., an ant colony) are composed of former grade 2 parts (multicellular animals), which in turn are composed of former grade 1 individuals (cells). Huxley uses the term "aggregate differentiation" (p. 48) to describe the hierarchical nestedness of systems that stay together and come to exhibit advanced internal division of labor. He identifies three hierarchical levels as originating by this process: molecules forming aggregates to become closed grade 1 cells, cells forming aggregates to become closed grade 2 multicellular organisms, and aggregates of what he calls "persons" aggregating to become closed grade 3 organisms (pp. 103–104).

Human condition: Huxley aims to minimize anthropomorphic bias in his analysis and, accordingly, the human condition is only explored tangentially. Huxley notes that our societies have the potential to advance in organizational complexity, but remain at a low level of individuality due to wasteful operating procedures.

The closure achieved by human cultural and political institutions remains incomplete and porous (pp. 108–109). Accordingly, Huxley seems to indicate that human societies are aggregations of grade 2 persons in the main body of the text. However, he appears to contradict this claim in Appendix A, where human society is listed as a higher organizational level. This tension, and its connection to Huxley's emphasis on the need for promoting harmony in human societies (p. 118), would shape his later career, but hardly affected *The Individual in the Animal Kingdom*.[1]

Individual: A closed organic system that is differentiated from its external environment, exhibits self-maintenance, and has an internal division of labor between heterogeneous parts that promote survival of the larger whole by enhancing its performance. Individuals also have adaptive features (pp. 11–12) and possess reproductive capabilities (pp. 20–21). Injury to component parts generally decreases their viability (pp. 36–37, 63). Cooperation of internal parts requires no coercion (p. 70), and the organismal whole is irreversibly more than the sum of the parts (p. 70).

Organism: Closely related to "biological individual" insofar as grades of individuality constitute discrete types of organismality. However, these terms are not strictly co-extensive. Individuality functions as a continuous variable, in the sense that more individuality correlates with quantitative increases in independence and heterogeneity of parts (pp. 8–9). In contrast, organismal grades are discrete, and include: unicellular (grade 1) and obligately multicellular organisms (grade 2), as well as individually or "person"-differentiated colonies (grade 3) (pp. 47, 91, 105, 117). Transitions between these grades are synonymous with MTEs.

Power of choice: An organism's naturally selected ability to actively influence its fate by gaining independence of the environment. Although specialization to narrow niches can reduce these powers (p. 100), they tend to increase over time within lineages and particularly across hierarchical grades (figure F.2). The modern term "agency" (in the sense of naturally selected behavior) is a close synonym (pp. 4, 116–117).

Protoplasm: The complete contents of a cell, consisting of at least three complementary substances: an outer layer (absorption/

protection), chromatin (assimilation/nucleus), and cytoplasm (catabolism) (pp. 44–45). Although sometimes regarded as internally homogenous by others of his era, Huxley was aware that protist cells are, in fact, internally heterogenous (pp. 13, 36–37, 67, 113). Protoplasm is described as self-regulating (p. 13), which implies that the individual wholes of all grades inform and define the functionality of their parts (pp. 73–74).

Race: A population of organisms, or sometimes the species as a whole (pp. 13–14). Occasionally used interchangeably with "individual" (p. 15) to indicate that species can be viewed as temporally continuous generations of the same basic type of protoplasm that predictably develops into a specific kind of organism.

Species: A set or type of organism that are able to reproduce fertile offspring. Generally used in a modern sense, but Huxley occasionally speaks of species when referring to biological kinds, e.g., of protoplasm (p. 43).

Teleology: Huxley accepts a naturalized form of goal directedness (*sensu* modern "teleonomy") that can be understood as an organism's adaptive drive to complete tasks (work) that promote survival and reproduction (p. 14). He suggests that understanding adaptation (pp. 43–44, 100) requires an integrated approach that appeals to naturalized teleology, materialism, comparative phylogenetic analyses (pp. 25–26, 38, 64, 117–118), and ontogenetic developmental studies (p. 28). This methodology was later formalized by Niko Tinbergen.[2]

NOTES

1. Krishna R. Dronamraju, *If I Am to Be Remembered: The Life and Work of Julian Huxley with Selected Correspondence* (Singapore: World Scientific, 1993).

2. N. Tinbergen, "On Aims and Methods of Ethology," *Zeitschrift für Tierpsychologie* 20 (1963): 410–433.

PREFACE

JULIAN S. HUXLEY

I must confess that when I made the choice of Animal Individuality as my subject, I had no idea of its real importance, its vastness and many ramifications: the teaching of philosophical biology is in England to-day somewhat of a Cinderella. The working out of the concept, full of interest as it was, brought also regret; a book of the size could have been—should have been—made from every twig and a stout octavo from the central trunk. This might not be; and the unavoidable compression must be pardoned. The general reader must imitate the Organic Individual (p. 19) and take unto himself wings of thought and conscious effort to skip across the unbridged gaps that perforce remain; with them to aid, I think he will find the stepping-stones not too far apart. The professed biologist must not cavil when he finds some merely general truth set dogmatically down as universal; in biology (still so empirical and tentative) there are always exceptions to the poor partial "Laws" we can formulate to-day. To have qualified every statement that needed qualification would have added much to the book's bulk without aiding the argument or being really more "scientific."

My indebtednesses are great. It will easily be seen how much I owe to M. Bergson, who, whether one agrees or no with his view, has given a stimulus (most valuable gift of all) to Biology and Philosophy alike. The various Oxford philosopher-friends who have helped to comb out the tangles of a zoologist's mind know how grateful I am to them: I will not name them here for fear my heresies be laid to their charge.

Certain criticisms have convinced me that some explanation of the scope of this book will here not be out of place. The task I have attempted in the following pages is a two-fold one. First, I have tried to frame a general definition of the Individual, sufficiently objective to permit of its application by the man of science, while at the same time admitted as accurate (though perhaps regarded as incomplete) by the philosopher. Secondly, I have tried to show in what ways Individuality, *as thus defined by me*, manifests itself in the Animal Kingdom.

I wish here to point out in general, that the failure of one of these aims does *not* preclude the success of the other; and, in particular, this:—it is possible that the philosophically-minded, will quarrel with my definition of the Individual (p. 20) as a "continuing whole with inter-dependent parts" (to put it at its baldest). But even if he denies that the definition applies to the Individual, he must, I think, admit that it does apply to something, and to something which plays a very important part in the organic world. He will, I believe, after reading the subsequent chapters, be brought to see that every living thing is in some way related to one of these systems, these continuing wholes; and that such wholes, though they may not in his eyes deserve the name of Individual, are yet sufficiently widespread and important to merit some title of their own.

Put in other words, the major portion of this book is devoted to showing that living matter always tends to group itself into these "closed, independent systems with harmonious parts." Though the closure is never complete, the independence never absolute, the harmony never perfect, yet systems

and tendency alike have real existence. Such systems I person-
ally believe can be identified with the Individuals treated of by
the philosopher, and I have tried to establish this belief. But
what's in a name? *The systems are there whatever we may choose to
call them*, and if I have shown that, I shall be content.

In conclusion, I will only hope that this little book may
help, however slightly, to decrease still further the gap (to-day
happily lessening) between Science, Philosophy, and the ideas
and interests of everyday life.

J. S. HUXLEY.
BALLIOL COLLEGE,
OXFORD.
Sept., 1912.

The numbers in brackets to be found in the text refer to the Bibliog-
raphy at the end of the book.

An Appendix has also been added, giving some of the main conclusions in tabular form.

EDITORS' NOTE

Huxley's original remarks indicate that cross-references are to be
found in brackets. In this edition, they are enclosed in parentheses.

I

THE IDEA OF INDIVIDUALITY

"Die Zeit ist abgeflossen, wo mir noch Zufälle begegnen durften; und was *könnte* jetzt noch zu mir fallen, was nicht schon mein Eigen wäre!"
NIETZSCHE.*

"La vie manifeste une recherche de l'individualité et tend à constituer des systèmes naturellement isolés, naturellement clos."
BERGSON.†

"Accidents cannot happen to me." So says Nietzsche's Zarathustra, and in the saying proclaims to the world the perfection of his individuality. It might be thought that such a being was far outside the purview of the Zoologist, that he himself belonged to imagination and his individuality to the most speculative philosophy, and that both he and it should be left where they belong, where they could not contaminate the "pure objective truth of science."

* Nietzsche, "The time is now past when accidents could befall me; and what could now fall to my lot which would not already be my own!"

† Bergson, "Life manifests a search for individuality and tends to form systems that are naturally isolated, naturally closed."

That I think is an error: for the idea of individuality is dealt with of necessity both by Science and by Philosophy, and in such a difficult subject it would be mistaken to reject any sources of help. Not only that, but animal individuality with the advent of consciousness, though still remaining a lawful subject of the Zoologist, becomes naturalised in the proper realms of the Psychologist and the Philosopher and transfers thither the major portion of its business.

More, even were the Zoologist to confine himself to a description of non-conscious organic individuals and the deductions he drew from them, he would often find himself without a reasoned criterion of Individuality or a true idea of what he means by "higher" or "lower" individualities. It is only when the Biologist and the Philosopher join hands that they can begin to see the subject in its entirety.

There are two chief ways of enquiry into the meaning of things—the static and the dynamic. In determining the nature of Individuality, for instance, we may seek to define it by comparing the different objects we are agreed upon to call individuals and then taking their Highest Common Measure—extracting from them the utmost which is common to all and erecting that as the minimum conception of Individuality; or we may search for the movement of individuality through the individuals, and, finding that some are more perfect, some more rudimentary in their individuality, thus establish a direction in which its movement is tending, and from that deduce the properties of the Perfect Individual, possessing then a maximum conception of Individuality.

In view of the change, the progressive change or evolution which is one of the fundamental things of Life, the second method is the more natural, and in a way includes the first. Using it in the main, therefore, but not rejecting the other as an engine, we will begin to lay siege to the notion of individuality; and so, having justified the necessity for some philosophical view of the subject, but with apologies none the less for a biologist's intrusion on another's domain, we return to Zarathustra and his pronouncement.

"Accidents do not happen to me."—When a glance is thrown over the various forms of animal life to which the name of Individual is naturally conceded,[1] it is seen that in spite of many side-ventures, they can be arranged in a single main series in which certain characters are manifested more clearly and more thoroughly at the top than at the bottom. One of these characters is independence of the outer world and all its influences—in other words, immunity from accidents. By independence is not meant the independence of the recluse or the ascetic, but that other independence belonging to the great man of action and the inventor. These are not independent in the most literal sense—they do not "do without," they are not proud of existing on the barest minimum; the ultimate logical end of that kind of independence is atrophy, both mental and physical. Their other, higher independence involves this much of dependence, that they employ the things of the external world as material with which to work. For the making of bricks, you are dependent upon straw: but you attain a higher independence by making bricks and being dependent upon straw than by being independent of straw and lacking bricks. They gain their independence by using the outer world for their own ends, harnessing some of its forces to strive with and overcome the rest. At the least they can resist the adverse current, displaying a purpose of their own which is not whirled away by every wind of fate. "Accidents cannot happen to me"—so spake Zarathustra, and then added this reason: "Because all that could now happen to me would be my own."

In this making of Nature his own, civilized man has an individuality vastly fuller, more perfect, than the savage. Both in resisting adverse forces and in harnessing the indifferent to his will, he is far superior; take as a concrete instance, for one the stamping out of malaria in the Suez Canal zone, and for the other the invention of the microscope.

At the other end of the series, even the simplest Protozoan has something of the same power. Although in a current against which the savage (let alone the steamboat of the civilised man) could easily swim, the Protozoan is carried utterly away, yet none the less it has some power of independent movement, and is not helpless like the inorganic grain of dust.

This gradual increase of independence up from the Protozoa to the highest animals is due partly to mere increase of size:[2] the same current that carries the grain of sand in its midst and rolls the pebble on its bed, swirls powerless past the boulder.

Partly it is due to increased complexity: the actions of the caterpillar who once in his life weaves an elaborate cradle to support his transmuted pupa-self, without either practice or the sight of another to teach him, can only be due to the actual machinery of his brain, working in a way almost as stereotyped as our machines,—a long series of ready-wound clockwork which must unwind itself when a certain catch is released. The Protozoan or the Jelly-fish is not capable of such precise and ordered action because it has not the requisite machinery, the requisite complication of brain and muscle.

Lastly it is due to increased adaptability, which depends mainly upon increased power of choice. Adaptability seems to be a property soon acquired by a complex and unstable substance, or rather mixture of substances, like protoplasm. Roux (16) by extending Darwin's idea of Natural Selection or survival of the fittest from individuals to the organs and tissues, the cells and varieties of protoplasm within the individual, has shown that some measure of adaptability, or useful response to changed conditions, becomes a common property of all living things. This, though very important, has been slow in action, merely automatic, and therefore limited in its usefulness, the result, to speak in metaphors, not of choice but of habit. What we call choice has only become fully realized through a special arrangement of special tissue—the brain.

Says Bergson: "A nervous system with neurons placed end to end in such wise that, at the extremity of each, manifold ways open in which manifold questions present themselves, is a veritable reservoir of indetermination" (1, p. 133). Such is the nervous system of man: and whatever value we assign to the idea of indetermination, whether we believe in the reality of choice and free-will, or think that they are only apparent, due to the relativity of our mental powers, the fact remains that in a brain which is constructed after the pattern of our own, and in which therefore we postulate the existence of Consciousness, a new machinery, different in kind from any machinery we have been able to construct, has been introduced; machinery that by supplying the individual with memory and reason gives him the largest scope to adjust his actions, and so himself, to the variations of circumstance.

Civilized man is the most independent, in our sense, of any animal: this he owes partly to his comparatively large size, more to his purely mechanical complexity of body and brain, giving him the possibility of many precise and separate actions, and most to the unique machinery of part of his brain which enables him to use his size and the smoothly-working machine-actions of his body in the most varied way.

But he is far from perfect independence of accidents. A being to whom accidents really could not happen might attain to that happy state through having perfected himself in any of the three qualities which have been seen to assist independence. By incorporating more and more matter—that is, by increasing in size—until co-extensive with the universe, he would obviously be entirely independent; there would remain nothing on which to be dependent. Since matter is what it is, man at least has little chance of advancing far along that road. By building up within himself a separate machine for dealing with each possible eventuality, independence would likewise be obtained were it not that there is an infinity of eventualities, and so the project is self-contradictory. But by perfecting his mental attributes—his means of perceiving, remembering, and reasoning—he would become capable of dealing with any one of the infinite eventualities, for though he could not construct an infinity of machines simultaneously, yet as each new eventuality cropped up, he would be able to invent a new plan to cope with it. Though Zarathustra had climbed far up this path, he probably was not quite accurate about the accidents: it is not likely that he would be able to experience everything, to remember everything, and to understand everything, but so alone would he be altogether immune from the accidental. That is neither here nor there. The chief importance lies in this: all life of which we have any assured cognizance is dependent upon or inseparably associated with a certain kind of matter—protoplasm. Knowing what we do of the properties of protoplasm, it becomes evident that no considerable advance towards independence through either of the first two methods is physically possible for life; it is only the third way, with its multiplication of potentialities, which, in spite of size really not so hugely great and mechanism really not so vastly complex, can yet give life a considerable fresh amount of immunity from accident.

The second quotation at the head of this chapter seems at first sight to take a very different view of the individual, conceiving of it as "a system naturally isolated, naturally closed." By this Bergson means that in any consideration of that system, it is the unity of it as a whole that is important: more than that, even if you want to consider a part of the system by itself, you cannot do so, for it loses almost all its significance when detached from the whole. What is the meaning of the hand and its actions apart from the functioning of the whole body? More striking still, for here there are no physical connections to sever, what is the meaning of a lonely bee and its actions when it comes back to find its hive destroyed? With inorganic things on the other hand, a part does not lose significance when detached from a system, nor the system appear less perfect for the detachment of the part. The inorganic system is a Particular, but not an Individual. Cause half a mountain to be removed and cast into the sea: what remains is still a mountain, though a different one. Take away a planet, and the Solar System still works: its working is different, but, as far as we can see, only different, not less perfect.

Nietzsche's words affirmed the individual's principle of action: Bergson's point out the inner unity for the good of which that action is performed. From the latter we can deduce another attribute of individuality—its heterogeneity; from that very unity of the whole we can postulate diversity of its parts. This sounds paradoxical, but in reality it can be easily shown that nothing homogeneous can be an individual.

Suppose (as is highly probable) that the earliest forms of life were homogeneous in chemical composition. If so, even were they compelled by the nature of things (see Chap. II) to exist as separate masses of defined shape and size, even though, by reason of their complicated atomic structure, they could

carry on all the diverse functions necessary for their continued
existence with their one chemical substance, they would then
not be individuals. There is no unity residing in such masses—
they are the merest aggregates; whether you divided one into
two or twenty or a hundred pieces it would still go on working in the same way, without a break,[3] whereas if you divide
a man into two by cutting off his hand, the working of the
main part—the man—is rendered less effective, and that of
the lesser part—the hand—is stopped for ever. Even in animals with the most astounding powers of regeneration, the
working of the whole is always impaired, if only for a short
time, by the removal of a part: some regulation, or remodelling, is necessary before the mutilated mass is ready to function
as a whole once more. Even such an animal is a whole and no
mere aggregate: it has an inner principle of unity, which may
be loosely fixed and lightly changed, but is none the less real.
Our hypothetical homogeneous masses have, in themselves,
no inner principle: their definiteness is imposed on them from
without, and one feels that if the external conditions altered,
they would have none of the independence of our perfect
individual, but would alter blindly with the conditions, like
raindrops, which in ordinary showers are small, but in a thun-
derstorm, under the influence of electricity, run together into
large heavy drops showing no sign of their composite origin.
One can, in fact, consider the working of any portion without
the slightest reference to a whole, and it thus becomes evident that nothing homogeneous can be called an individual.
Starting from the just not homogeneous, there can be traced
a tendency towards ever greater heterogeneity running up
through the series of animal individuals. This was indeed only
to be expected. To perfect its independence, the individual,
it was seen, had to render its actions precise, independent of
each other: and in non-conscious organisms at least, difference

of function always implies difference of structure, so that the more independence—the more individuality—an individual is to possess depends very closely on the amount of heterogeneity of its parts. Look for instance at such an individual as a colony of Termites ("white ants") (cf. p. 108), its defence delegated to one caste, its nutrition to another, its reproduction to another; the various castes are specially adapted in their structure for their various functions. It is obvious at once that the queen with her vast swollen abdomen full of eggs is a much more effective reproducer than if she had retained any of the structure and mobility necessary to defend or look after herself. The soldiers again could not have been such powerful defenders of the colony if they were to have kept any of the delicacy of mandible required by the workers, the craftsmen.

Another illustration: the accurate grasping powers of the human hand are only rendered possible by its consisting of a number of distinct but co-ordinated parts. The action of grasping is an undivided and a single act, but is only possible because the organ of grasping consists of separate and different parts. The pseudopod of an Amoeba, to take the opposite extreme, has no differentiation of parts: hence the functions it can perform are few and unprecise.

In both these cases, the dependence of efficient action, and so of independence, is clearly dependent upon a visible and obvious heterogeneity of structure. It might appear self-evident that the organs, the animal's living tools, should have a different structure according to the functions they were meant to carry out, were it not that in man we have the example prominently before our eyes of an enormous number of very special functions being executed by a single organ such as the hand. This apparent exception is due to the structure of his brain, which has given him reason and educability for instinct and automatism. True, he has to be at the trouble of exercising his wits, but gains vast potentialities thereby; the brutes have no toils of learning, but their smooth actions are sadly limited. He has learned to make tools from inorganic materials, and *they* serve as the heterogeneous structures by means of which he can perform all his diverse actions. For specialised functions there must always exist specialised structures; but man through his conscious reason has been able to put off the burden of them from his own substance on to the broader shoulders of inorganic nature. There does exist some corresponding heterogeneity in himself, but not in visible structure: it lies in the diversity of his states of consciousness.

These cannot all exist as such at one time,[4] but by means of the memory, each can be summoned up as it is wanted. No doubt accompanying them there are physical and chemical differences in the nervous tissue, causing differences of continuity between the various neurones, but this physical heterogeneity is of no obvious or visible kind. The broad differences, the differences that can be felt, lie in the states of consciousness, so that the individual, after advancing a long way in its march towards perfect individuality by means of heterogeneity of co-existent structures, has got to its present position by adding to this a new device, heterogeneity of states of consciousness, which states, through not being co-existent, can be more numerous and more heterogeneous than ever the structures could.

One last attribute of the individual, but a very important one. So far the individual has emerged as "Unity in Diversity." It shows diversity both in what it *is*—its physical structure and the architecture of its consciousness—and in what it *does*—the actions which more truly constitute its real essence. It also has unity, because though all its heterogeneity of architecture is devoted to producing heterogeneity of actions each one of these only has meaning when considered in relation to the whole. Thus the problem so far has been the relation of the parts to the whole. There remains to be considered the relation of this whole to itself.

Since it is obviously the working, the function, which is important in an individual, the structures being only instruments for the function's better performance, this question really resolves itself into the relation between the working of the whole individual at one time and its working at another, later time. This has already been implicitly answered. When we said that the hand and its functioning had significance in relation to a whole, we did not mean merely to a whole which happened to be there at that one instant, but to a whole which had a continued existence in time. When the hand takes up a piece of bread and puts it into the mouth, that action has no significance for the whole man if only that instant of time is considered. Its significance is only seen later, when the bread has been digested, absorbed, and carried to nourish all the hungry parts of the whole individual.

What has been said so far presupposes some degree of continuance in the individual; a survey of the various kinds of organic individuals shows this continuance to be common to them all, and that too in no limited measure, but as one of the fundamentals of their existence. Looked at from this point of view, the individual appears as a machine whose working has for result no "finished article," the uses of which do not affect the machine, but merely the continuation of that same working. The result (and the object) of the working of a printing-press is to print books: but the books when printed are of no use to the press. The result (it is risky to say the object) of the working of an individual is for it a minute later to be still working in the same way. There is no material product given birth to by the process; but the result of the working is of the greatest interest to the individual, the machine that is working.

This fourth view of the individual, as a whole whose diverse parts all work together in such a way as to ensure the whole's continuance, or, as the evolutionist would say, whose structure and working has "survival-value," cannot stand without some qualification. There is death to be reckoned with; the survival is only temporary.

Under cover of the one word Death lie sheltered two separate notions—death of the substance, when the living protoplasm ceases to exist as such, and death of the individuality informing the substance.[5] In man, both are inseparably connected; in many lower animals they are not. To take the simplest example: most Protozoa, such as Amoeba or Paramaecium, definite individuals both, feed and continually grow, and when they are grown to a certain maximum size, divide into two halves (see pp. 33, 42–43), each of which reorganizes itself into an individual resembling its "parent." Not a jot of substance has been lost: but one individuality has disappeared and two new ones are there in its place.

Owing to the material properties and limitations of her "physical basis" of protoplasm, Life in her attempt at perfect individuation has been faced by a dilemma with which she has never fully been able to cope.

Growth, the balance of gain over loss in metabolism, is either a necessary attribute of protoplasm, or else, more probably, an easily-acquired property, of such all-round usefulness that every organism has seized upon it (see Roux, 16). At all events it is universal in all protoplasm throughout all or most of its active existence. Now if Life allows this growth to take place indefinitely within the limits of one individual, two awkward things happen; first of all, the mere increase of bulk brings difficulties (see Chap. II), and secondly the increased weight of the whole needs some kind of a skeleton or scaffolding for its support. This skeleton, since living protoplasm itself is not firm enough, must be built out of dead materials, mere secretions of protoplasm. These have not Life's power of renewing themselves, of "sprouting fresh and sweet continually out of themselves" like protoplasm, yet all the time are being exposed to the inclemencies of the world and the assaults of enemies: at last something, the oldest part, gives, and involves the whole fabric in its fall.[6]

Death of the substance—that has been the result whenever Life has allowed unlimited growth to the individual: and when she preserves the substance, as in the Protozoa, by dividing it into two whenever it has reached a certain size, so keeping the pattern of the race within a narrow range, easily controlled, then there must be death of the individuality. She has never been able to produce an individuality which can for ever keep the unstable structure of its substance nicely balanced against the chance violences of the outer world.

But—and this is important—when the Protozoan divided its substance and destroyed its individuality, two fresh ones sprang up in the two separate masses of substance. The relation of the organism's individuality to its substance will be considered at more length in Chap. VI. Here it can only be said that protoplasm has primitively a great power of self-regulation, so that the plan of the individual's structure which is characteristic for the species can exist actual and patent in a given mass of protoplasm, and yet can also exist, though latent and potential only, in any and every part of that mass above a certain minimum size. Break off a Begonia leaf and chop it into little bits; each bit reveals its latent power, sending roots downward, shoots upwards, and at the last becoming a self-sufficient whole. Through this regulatory power, Life has been able to save herself a tossing from her dilemma, escaping, like a Minoan acrobat, between the very horns: through it she has the possibility of reproduction.

The essence of reproduction is that one individual should create a new individual out of itself. The parent may persist, as in man, after the offspring has come into the world, or, as in Protozoa, may annihilate itself in the very act; that does not matter. What matters is that in every species there exists a succession of individuals in time, each one derived from the very substance of an earlier, each one built up and working on a common plan. Life has thus been able to steer a middle course. In the higher animals, for instance, she has perfected and used the single individual up to a point, to procure the greatest amount of independence for herself who animates his frame; then, when it becomes difficult, and more difficult as time goes on, to maintain his supporting tissues in repair and hold balanced the many processes struggling within him, she calls in the power of reproduction, raises up new individuals of the same sort out of his substance, and abandons him to his fate; but the race goes on.[7]

Our first definition of the individual based on the idea of continuance can now be amended. We must not say that the individual is a whole whose parts work together in such a way as to ensure that this whole, and its working, shall persist; the individual only persists for a limited time. In spite of this, something does indefinitely continue, though it is but the kind, the species, and not the single individual itself. There is only one kind of working in the species, and this repeats itself in a recurrent cycle; but for each cycle as it recurs a new individual is required as the instrument of the working.[8]

These qualifications, universally applicable though they are to all individuals that we know on this earth, are still mere qualifications, not essential to the pure idea of individuality: the perfect individual would be eternal, subduer of time as well as of space. Since, through practical difficulties, Life has not been able to reach this perfection, she has had to content herself with the next best, continuance of the kind of individual instead of the individual itself.

This, however, is alone enough to rule out of court the pretensions of all inorganic constellations to individuality, those even of crystals and of solar systems. The solar system is a whole most definitely "isolated by Nature," heterogeneous, and composed of parts closely inter-related in their working; what, besides the objection made above (pp. 6–7), which may only depend on our ignorance, prevents our calling it an individual? This, that its working is not directed to continuing either itself or other systems like itself.

The crystal has no parts, but is homogeneous; were it not, its working would still betray it, though at first sight its growth and its strange powers of regeneration display it as functioning to preserve a special form. Put in a weak instead of a saturated solution, and it will not simply cease to exist, like an animal placed in unfavourable conditions, but will unbuild itself as busily and regularly as just now it built itself up. Such a combination of two diametrically opposed and equally active tendencies can scarcely be called an individual.

The existence of a species or race, a procession of similar individuals each descended from a previous one, as well as of what we usually call individuals, the separate beings that at any one moment represent the species, leads of necessity to the separation of two distinct kinds of individuality, one belonging to the race and one to the persons that constitute the race. Take as an example *Distomum hepaticum*, the Liver Fluke (Fig. 1). The eggs of this unpleasant creature, which gives sheep the disease known as liver-rot, are passed out of the host and hatch out into minute embryos that swim about in the film of moisture on the meadow-plants. They cannot develop further unless they fall in with a particular sort of snail: if so, they burrow into its liver, and grow up, not into a new fluke, but into an irregular sort of bladder, the *sporocyst*; this, from its inner wall, produces a number of new embryos which grow and burst out of their parent as the so-called *rediae*—individuals differing both from the fluke or the sporocyst. These in their turn give rise to a number of little tailed creatures, the *cercariae*, which migrate out of the snail, pass into a resting stage on blades of grass, and there passively await a browsing sheep. If one by good chance devours them, they hatch out, bore their way into the liver, and grow up again into flukes.

Now each of these three forms that thus cyclically recur is obviously an individual in the sense defined by us: they are wholes with diverse parts, whose working tends to their own continuance, even though this continuance is limited. But besides this there is the cycle itself to be reckoned with: it too is a definite something, a whole, it too is composed of diverse parts, sporocyst, redia, fluke, it too works in such a way that it continues (and continues indefinitely). What right have we to deny it an individuality as real as those possessed by any of its parts? True, those parts are separated in space; but the ant-colony (p. 108) shows that this is no bar to individuality. The real point is this: the existence of the sporocyst and the redia is of no *direct* advantage to the individual fluke: it would grow and lay eggs just as happily if all the host-snails, and with them all sporocysts and rediae, present and to come,

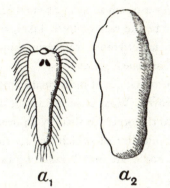

Figure 1
Diagram of the life-history of the Liver-fluke. The egg hatches out
into a free-swimming embryo (a_1); this if it finds its snail changes into a
sporocyst (a_2); this produces inside itself a number of rediae (b); which
in their turn each produces a number of cercariae (c_1). These, if
conditions are favourable, find their way into a sheep, where they grow
up into the adult Fluke (c_2) (a_1—c_1, magnified; c_2 natural size.)

were exterminated. It is however of advantage to something, and that something can only be the race of liver-flukes, the kind of protoplasm which by its difference from other kinds has earned a special name—*Distomum hepaticum.*

That is an extreme case; the two kinds of individuality may often be inextricably interwoven. What is of advantage to one is usually of advantage to the other, so that, by an over-emphasis of the species-individuality of which we are the parts, it is often said that our bodies are only "cradles for our germ-cells."

It must here suffice to say that wherever a recurring cycle exists (and that is in every form of life) there must be a kind of individuality consisting of diverse but mutually helpful parts succeeding each other in time, as opposed to the kind of individuality whose parts are all co-existent: the first constitutes what I shall call species-individuality, or individuality in time, while the other corresponds to our ordinary notions of individuality and, if a special term is needed, may be called simultaneous or spatial individuality. It is of individuals of this latter class that we have so far been speaking, and to them we must now return.

Our minimum conception of continuance—the continuance of the kind of individual rather than of the single individuals themselves—is thus a touchstone to distinguish between what is and what is not an individual: it now remains to trace the progress of continuance on this earth up towards the unattainable maximum of the undying. At the start, the individual in such organisms as bacteria has a duration reckoned merely in hours or even in minutes. There is but the hastiest procession of never-returning forms across the stage of the species. As we ascend the scale, the individual learns to stay longer and expound his part more clearly. With the attainment of the multicellular condition and the possibility of reproduction by detaching one small part of himself instead of by division of the whole (p. 35), he can even linger on the stage till the next scene is half played through.

In the actual duration of his life, the individual ranges from the bacterium's hour to the big tree's five thousand years. So far the direct and obvious path can lead. But consciousness once more has found out a way more subtle and more effective. Man in this again stands on the pinnacle of individuality—not in mere length of days, but in having found a means to perpetuate part of himself in spite of death. By speech first, but far more by writing, and more again by printing, man has been able to put something of himself beyond death. In tradition and in books an integral part of the individual persists, and a part which still works and is active, for it can influence the minds and actions of other individuals in different places and at different times: a row of black marks on a page can move a man to tears, though the bones of him that wrote it are long ago crumbled to dust. In truth, the whole of the progress of civilization is based on this power. Once more the upward progress of terrestrial life towards individuality has found apparently insurmountable obstacles, gross material difficulties before it, but once more through consciousness it finds wings, and, laughing at matter, flies over lightly where it could not climb.

One word more on continuance. The continuance of the working of a species as we have defined it would preclude change; but change and the idea of evolution are at the base of all modern thought in science and philosophy alike. As a matter of fact, the resemblance of the working of one individual to its result, the working of a descendant individual, is never absolute: and so, since working and structure are inter-dependent, no two individuals are ever exactly alike in appearance and architecture. Given this fundamental fact of variation, nothing is impossible: and to-day few would be found to deny that all the battalions of living organisms are descended from one primeval type. That is the logical outcome of the doctrine of Evolution. Evolution is a word glibly used, but often without thought of its full meaning. If Evolution has taken place, then species are no more constant or permanent than individuals. We know what we mean when we use the words child and man, and we know that at puberty comes the crisis which transforms the one into the other; but the whole process is continuous. So we know what we mean by a species; probably, too, there are crises when the species becomes unstable and in a short time we can say, "here is a new species." None the less the one species, if we accept the idea of Evolution, is continuous with the other by the most obvious continuity, that of its substance. As individual emerges from individual along the line of species, so does species emerge from species along the line of life, and every animal and plant, in spite of its separateness and individuality, is only a part of the single, continuous, advancing flow of protoplasm that is invading and subduing the passive but stubborn stuff of the inorganic.

From this short survey of the types and tendencies of existing individuality, three things emerge. First comes the minimum conception of an individual; the individual must have heterogeneous parts, whose function only gains full significance when considered in relation to the whole; it must have some independence of the forces of inorganic nature; and it must work, and work after such a fashion that it, or a new individual formed from part of its substance, continues able to work in a similar way.

Then comes the idea of the perfect individual—something unknown to our senses, its characters a mere raising to infinity of those enumerated above. Defining those characters in different form, we may say that such a being would possess perfect internal harmony, and perfect independence (in our particular sense) of matter and of time itself.

Lastly, and this is perhaps most important for the present quest, there shows the actual line traced by Life in her progress up towards this perfect individuality. She has had to contend with the limitations of her own physical basis, and the result achieved is a compromise; not what she planned, but what her imperfect materials allowed her to carry out—the old difference between the poem flashed on the poet's brain and the same poem on paper, striving to gleam through the words that build it.

Her track is straightforward at first: she tries to realize to the full the possibilities of her material basis, increasing the mere size, the mechanical complexity, and the length of life of her individuals, but at last there comes a point where she can go no further forward—the spirit is willing, but the flesh is weak. So far, range of action has been dependent upon actual mass of substance, diverseness of action upon complexity of substance, and length of action upon duration of substance. Now this direct way is barred: but she finds out another path. She produces a unique type of mechanism, of which the most fully developed type is the human brain, and, associated with it, the power of conscious reason and of memory. At once the individuality is released from waiting servile upon substance. Now to its own size it can add the size of all its tools and machines—by them now is measured the Range of its action:

the Diversity of its action it has multiplied a hundredfold by substituting indefinite potentialities for necessarily limited actualities; and the Duration of its action, by the device of language, now far surpasses the allotted span of its substance.

To such an individuality, one that can thus transcend the limits of its substance, the name of Personality is commonly given. Man alone possesses true personality, though there is as it were an aspiration towards it visible among the higher vertebrates, stirring their placid automatism with airs of consciousness. In man, personality is usually defined with reference to self-consciousness rather than to individuality; but the power of reflection and self-knowledge is linked up, in our one type of personality at least, with the new flight of the individuality—conscious memory seems necessarily to imply a vast increase of independence, so that it is all one whether we define the possessor of a personality as a self-conscious individual, or as an individual whose individuality is more extensive both in space and time than the material substance of its body.

Personality, as we know it, is free compared with the individuality of the lower animals but it is still weighted with a body. There may be personalities which have not merely transcended substance but are rid of it altogether: in all ages the theologian and the mystic have told of such "disembodied spirits," postulated by the one, felt by the other, and now the psychical investigator with his automatic writing and his cross-correspondences is seeking to give us rigorous demonstration of them.

If such actually exist, they crown Life's progress; she has started as mere substance without individuality, has next gained an individuality co-extensive with her substance, then an individuality still tied to substance but transcending it in all

directions, and finally become an individuality without substance, free and untrammelled.

That for the present must be mere speculation. The Zoologist has strayed: he must return to his mutton and his amoebae,
and in the next chapter will begin to consider more closely
the actual facts of animal individuality and their probable
explanation.

II
———

THE BIOLOGICAL FOUNDATIONS OF
INDIVIDUALITY

The idea of individuality, in common with most other large biological problems, came to be first considered—as indeed was only natural—from the standpoint of man alone. With the growth of our knowledge concerning invertebrate animals, the ideas thus gained had to be considerably modified, until finally the theory of evolution once and for all justified the more advanced among the earlier thinkers, and showed that in any view of animal individuality as a whole, we must not take man and mammals as the single starting-point whence we could logically work backwards to all the rest of the organic world, but must regard them as an ending instead of a beginning, and, what is more, as but one ending among many. From the single beginning, many lines have branched out to the many endings, and the only logical method is to start from the beginning (where, too, the phenomena themselves are far less complicated) and trace out each line to its ending, instead of trying to bring the various endings into relation with each other. Each ending is only intelligible through its history, and the history of one is different from the history of another.

The one advantage possessed by the anthropomorphic view of individuality (which, as a half-unconscious product of every-day experience, is still held by the great majority of those who are not professed biologists) lies in its dealing with long-familiar things. Since, however, this is a very real advantage to those who are approaching a subject the major part of which is bound to be not at all familiar, "full of strange oaths," and so bristling with new names that "bearded like the pard" is scarce a stretch of metaphor, we shall begin here with man; thence, taking stock of the more obvious facts of comparative anatomy, with the historical or evolutionary idea to aid us, try to extend the conception from man to the rest of the animal kingdom; then we shall have to show some of the chief difficulties which attend upon this point of view; and finally, having thus cleared the decks for action, we shall be able to take up our subject anew from its historical and logical beginnings.

A normal adult man or woman is an organism, whose complicated and varied parts are almost all designed for one end—to prolong the existence of the whole to which they belong.[1] It is in fact a machine which has the power of running itself, independent, within wide limits, of what is happening in the rest of the world. Unlike our artificial machines, however, whose working is constant, and whose only change is one of wearing-down, the running of the organic machine leads to changes in the actual structure of the machine, and so to changes in its working. We develop of necessity, of necessity we age, and at the last we die. But we remain the same individual throughout—on that all common use is agreed. Till death, when we obviously cease to be whatever we have been before, we preserve our individuality in spite of all fundamental differences in appearance and behaviour. But as to our nature before birth, there the common view is at a loss; its uncertainty has found expression in Milton's words, when to Limbo he consigns, not "Eremites and Friars" only, exiled thither for theological reasons, but "Embryos and Idiots" as well.

The very conjunction of his words will help us out of the difficulty. In our thought, the idea of human individuality has become interwoven with that of personality—a purely mental attribute. Even though by *embryo* Milton meant *abortion*, the lack of mentality—of personality and of soul, if you will, which it shares with the idiot, is the same whether it be within or without the womb, and he was right in regarding its fate as a grave theological problem. But (though the reasons for the defect of mental power are different in the two cases) an embryo cannot because it lacks personality be considered to lack individuality too, any more than an adult idiot can, although the individuality is no doubt less intense or perfect than in the normal adult man. It is this confusion of personality and individuality that raises most doubts in the mind of the average man as to the claims of the foetus to be called an individual. The other chief doubt arises from its incapacity to live out of its mother's body. But reflection will show that the embryo is like every other living thing in being able to exist only under certain defined conditions, which are merely much narrower for it than for the adult man and the generality of animals. (See pp. 100–101.)

Thus we can take the individuality of man back before birth to a stage when the embryo ceases to be easily recognized by the naked eye. To trace it still further, the man of science with his microscope and his knowledge of simpler animals must step in. There the ordinary man must pause, and there we will leave the question for the present; turning now to see how far his anthropocentric notions of individuality radiate out to other living things.

We find that he unquestioningly applies the word to all the familiar creatures of everyday acquaintance, the four-legged beasts and the birds, the snakes and the fishes. This is his

unconscious Comparative Anatomy—he recognizes instinc-
tively the community of general plan he shares with them.
This unconscious reasoning will carry him still further: he will
not hesitate when it comes to snails or insects or worms—in
fact, show him anything with a mouth and a stomach and he
will dub it an individual. So far, all seems plain sailing.

In reality, this is exactly where all the difficulties begin:
without studying the outward form and minute structure of
himself and of other animals at all stages of development—
without some knowledge, that is to say, of all the numerous
branches of the science of Zoology,—it is impossible for him
to extend his knowledge of individuality any further, and
when he does call Zoology to his aid, he finds that in every
direction it seems to lead to contradictions, raising difficulties
worse than those it lays. To start with, he has been considering
till now only those of the lower animals which slip naturally
into a scheme taken from the pattern of Man. A mustering of
all the clans soon reveals numerous types of animals that will
not fit this frame at all.

There are communities, such as those of bees and ants,
where, though no continuity of substance exists between the
members, yet all work for the whole and not for themselves,
and each is doomed to death if separated from the society of
the rest.

There are colonies, such as those of corals or of Hydroid
polyps, where a number of animals, each of which by itself
would unhesitatingly be called an individual, are found to be
organically connected, so that the living substance of one is
continuous with that of all the rest. Sometimes these apparent individuals differ among themselves and their energies are
directed not to their own particular needs, but to the good of
the colony as a whole. Which is the individual now?

Histology then takes up the tale, and shows that the majority of animals, including man, our primal type of individuality, are built up of a number of units, the so-called *cells*. Some of these have considerable independence, and it soon is forced upon us that they stand in much the same general relation to the whole man as do the individuals of a colony of coral polyps, or better of Siphonophora (pp. 89, 91), to the whole colony. This conclusion becomes strengthened when we find that there exist a great number of free-living animals, the Protozoa, including all the simplest forms known, which correspond in all essentials, save their separate and independent existence, with the units building up the body of man: both, in fact, are cells, but while the one seems to have an obvious individuality, what are we to say of the other?

So far we have treated the problem statically, as it were: when we come to view it dynamically, tracing the movement of life along its course, the difficulties do but increase. Take, to begin with, a simple colony of Hydroid polyps (Fig. 2), and ask how does this multiplicity of connected animals arise? Observation shows the whole stock to be formed, by a process of budding, from one original individual. A little lump or knob is seen at one place, which, growing rapidly, bit by bit assumes the appearance of the individual whence it has sprung; it takes its origin in a small group of cells (not in a single one) and its growth depends on continued growth of the substance of these cells, accompanied by their repeated division. By this means, the first individual produces a second out of itself. Its own individuality is not lost in the process; it is, however, impaired, for though the creature's organization is practically the same as it was before, yet it is no longer separate in space, and that part of it below the bud's point of origin is now the common property of the two individuals.

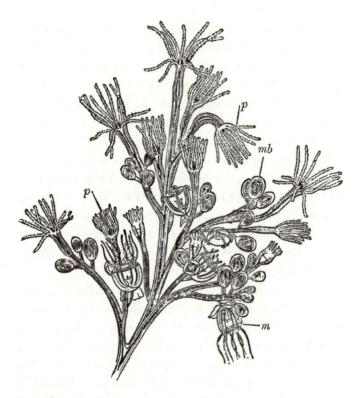

Figure 2
Portion of colony of *Bougainvillea fruticosa*, magnified. *p*, polyps; *m*, medusae; *mb*, medusa-buds. (From Lubbock, after Allman.)

The second individual and the remaining members of the colony, which are all formed in the same way, differ from their original only as regards their mode of development and, as a consequence of this, in never having enjoyed a free and full individuality. The relation of the individuals in a colony to each other is thus rendered still more obscure owing to the fact of one being produced out of another. What was at first nothing but a part grows up into a new whole.

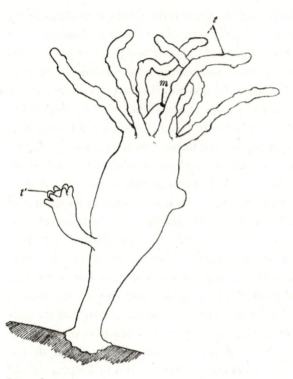

Figure 3
Hydra, semi-diagrammatic, showing a bud-rudiment on the right and
an advanced bud on the left. *m*, mouth; *t*, tentacles; *t′*, tentacles of bud.
(Magnified.)

Budding, though perhaps most striking when it leads to the
formation of a colony, is by no means restricted to colonial
forms: often, as in Hydra (Fig. 3), the process is completed,
and the bud set free to lead an independent life. Here one individual has produced a second out of its own substance: the
two resemble each other not less closely than two individuals
bred from the egg, and yet the first has lost not a jot of its own
individuality in thus creating itself anew in the second.

This fresh creation of new forms from the substance of the old is what we usually term Reproduction. Budding is but one of its many methods, and we must look at some others before we can see its full bearing upon our subject. First we will take fission, or division into two halves, a method which occurs in several groups of the higher animals, though less commonly than budding. Rarely, as among the stony corals, are colonies produced through its means; usually the two halves part company and each becomes as perfect an individual as its parent. It is, however, in this relation of parent to offspring that division is at variance with budding. Instead of one individual producing another, here the founder of the race ceases to exist, losing his own individuality in the production of two fresh ones. A glance at Fig. 4 will show that the whole substance and the whole organization of the first individual is separated in division into two discrete masses, each of which is incomplete in possessing only half the normal structure. These incomplete individuals, in the examples we have chosen, and in many other animals as well, do not as one might expect complete themselves by keeping the old half-organization intact and budding out what is missing, but, by a method involving a more radical destruction of the parent's individuality, they remodel their structure by a strange internal mason's-work, turning the materials that but now constituted a half-individual into a whole. From their parent they receive the half of its substance and the half of its organization; they make a new organization without adding to the substance.[2] Growth subsequently increases their size without altering their individuality or organization, until, on attaining to the prescribed limit, they repeat the process. Division is thus even more important for the present purpose than budding; we have the strange paradox that though each individual

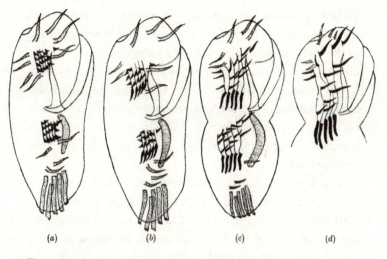

Figure 4
Four successive stages in the division of *Stylonychia mytilus*, a
Hypotrichous Ciliate Protozoan, from the ventral surface. (Highly
magnified.) The outlines of the mouth and the 17 ventral *cirrhi*
(locomoting bristles) are shown. Just before division there appear 17
small processes (drawn black) arranged close together in six parallel
rows. These enlarge, spread over the body, and become the cirrhi of
the daughter-cells, while the parent's cirrhi (stippled) become displaced
and at last completely absorbed. The new mouth-region is shaded
transversely. In (*d*), division is not complete, but only one half is
drawn. (After Wallengren.)

hands on the whole of its substance intact to its successors yet
with this perfect continuity of substance there co-exists perfect discontinuity of individualities.

There remains the third chief mode of reproduction. In
considering the hydroid colony, we found that all its members
took their origin, by budding, from one single founder. This
founder, though identical in organization with the rest, has
yet not had the same origin as they. Tracing its life backward

towards its source, we find it first of smaller size; then comes the stage when its organs are developed one by one, much as in the bud; before this it exists in a form through which the budded individuals never pass—as a small drawn-out ovoid, actively swimming instead of fixed to the ground; before this again it is seen as a round motionless body, built up like a mulberry out of rounded parts, and finally its "fount of life" is revealed in a spherical, inert mass, single and undivided—the fertilized egg.

This fertilized egg is neither more nor less than a cell—specialized, as one would expect, for the discharge of its own particular duties, but still a cell. Here is a further strengthening of our view of the higher animal or metazoan as a colony of units each comparable with a protozoan.[3] When the method is traced by which the plurality of cells in the adult arises from the single cell of the egg—the method, that is, of cell-reproduction—it is found to be identical with one of the ways of reproduction in metazoan individuals, that of fission; the single founder of the cell-community, the egg, divides the whole of its substance into two halves, each of which is a new cell. This is repeated again and again, and the whole army of cells in the full-sized hydroid are direct descendants of that single founder-cell. But the hydroid is itself a founder; and the new "individuals" which it buds out depend for their growth upon this same process of cell-division continually repeated.

The paradox is growing yet. Each hydroid seems in its way a whole; yet it is as well a mere part of a single greater whole, the colony, and, besides this, itself composed of units each of which again is in some sort a whole: and each whole has some claim to the name of Individual.

One gap still yawns: what was the origin of the single cell that gave birth to the whole adult organism? In this particular case, it was a fertilized ovum: by which is implied that the single cell has arisen from the total fusion, body and soul, or rather cytoplasm and nucleus, of two other cells, these are technically known as the *gametes*, and their product, the fertilized ovum, as the *zygote*. These two cells have come from two separate individual persons (one male and one female) and their cell-ancestors have been firmly built into the fabric of those individuals' bodies.

This merging of two cells and their two individualities in one (the exact reverse of fission) is the essential sexual act, and is usually known as the *conjugation* of the two cells. It will be considered more fully later (p. 54); here it does not concern us, for, as Weismann and others have conclusively proved, reproduction and conjugation are in their origin totally distinct from each other. In all the Metazoa, however, conjugation is always connected with reproduction, so that the fusion of two cells always implies the production of a new individual. In ourselves, and all other Vertebrates, the converse also is true, that the production of a new individual always implies the previous fusion of two cells—reproduction, in other words, is always sexual: but in very many of the lower Metazoa, though conjugation leads to reproduction, reproduction may occur independently of conjugation. Two examples of this asexual reproduction have been seen in budding and in fission.

Thus the complication introduced by the fusion of the two gamete-cells into the otherwise unvaried succession of cell-divisions does not really affect the present question, the relation of one individual to another. The essential point lies in the continuity of individual with individual.

To add the final straw, regeneration comes. Regeneration is usually looked on as something strange, almost abnormal, owing to its not occurring in man or his animal familiars. In reality it is much rather an original property of life, which for special reasons has dropped out of the human scheme of things.

As we descend the vertebrate scale, it is not until we reach the lower Amphibia, such as the newt and salamander, that regeneration becomes at all marked. Even here it is present in a restricted form, and is confined to the restoring of lost organs. A leg, that is to say, or a tail, even an eye or a jaw may be replaced, but the central systems and main lines of organization must be left intact. There must remain a certain central residue of the individual if it is to complete itself.

This in itself points to a vaguer, more fluid notion of individuality than can ever be got from contemplation of man alone, but what are we to say of such things as happen in many of the lower animals? Take first one form of regeneration seen in Clavellina, one of those poor relations of Vertebrates, the Ascidians. Cut Clavellina across in the middle, and (in certain defined conditions) a bud will sprout from the front end of the hinder half, and another from the hinder end of the front half. As in the growth of the hydroid colony, the old organization is kept entire and whole, the new organization is built up in the bud; here, however, it is not one whole individual giving rise to another, but one half giving rise to just that dissimilar half which is its complement. What is more, both the unlike halves of the original whole can thus add what is wanting— that and no more—to bring them up to the rank of wholes again. There is an old school-boy question about a cricket bat:—suppose the handle of a bat broke, and a new one was put on to the old blade. Suppose then that the blade broke and was in its turn replaced; would the bat still be the same bat? That is a hard question, but Clavellina asks a harder still.

From this to the extremes of regeneration, such as occur among flatworms and protozoa, is another large step. Stentor, for instance (a protozoan which happens to be specially convenient for experimental purposes), may be chopped, broken, or shaken up into pieces of all sizes and shapes, and every piece, provided only that it is above a definite minimum size (less than 1/300 inch in diameter, and in bulk only 1 or 2 per cent. of a full-grown Stentor), and that it contains a piece of the nucleus, will blossom out as a minute but full-formed individual, which will feed and grow and be indistinguishable from a product of natural generation.

By now, all faith in man as a guide to individuality must have been shattered. In man, an individuality presents itself as something definite and separate from all others, something which animates a particular mass of matter and is inflexibly associated with it, appearing when it appears and vanishing only when it dies. That idea of individuality is not universally applicable.

In perplexing procession before us there have appeared individualities inhabiting single cells, others inhabiting single cells at the start, many cells (and each of these with some kind of separate inhabitant of its own) in later life: individualities whose fleshly mansions are continuous one with another, no boundaries between: individualities that appear and disappear along an undying stream of substance, the substance moulding itself to each as the water of a stream is moulded in turn to each hollow of its bed: within one individuality others infinite in number, lying hid under the magic cloak of potentiality, but each ready to spring out as if from nowhere should occasion offer.

Nothing remains but to abandon preconceived ideas. We must seek to interpret human individuality not as the one true pattern to which all others must conform, but as something with a history and intelligible only through that history. We must therefore make for the first beginnings of things and trace their upward progress. For this to be adequately done, the very fundamentals must be explored, and the quest begin with an enquiry into the original and essential properties of living substance.

The biologist, looking at life objectively, finds life then manifest itself as the sum of the properties pertaining to a group of peculiar and complicated chemical bodies which are classed together under the general name of protoplasm.

The form and structure adopted by the lowliest living things at the time of their origin, which then had to serve as the starting-point for all subsequent forms and structures in life, are chiefly due to two properties of these protoplasmic substances—one physical, the other chemical. The first is their colloidal nature, which permits of their sharing the definiteness and resistance of solids with the mobility and quick chemical reactions of liquids. The second is their power of assimilation, their power of building up, out of materials different from and chemically simpler than their own substance, new molecules, identical in composition with the old. Assimilation is molecular reproduction, and is by far the most important property of protoplasm. Whenever an organism performs any action, it must needs do work, expend energy. This energy it procures from the break-down and combination with oxygen of some of the unstable living molecules. Combustion is here associated with chemical decomposition: the result is not mere oxidized protoplasm, not protoplasm at all, but various more stable and more oxidized compounds. Every action thus necessitates the destruction of some of the living substance, and were it not for the assimilatory power, whereby it can pick up materials from the outer world and force them to assume a structure and arrangement like its own, all protoplasm would soon vanish into nothingness.

From these two fundamental properties of protoplasm we can understand three important and almost universal qualities of living things.[4] First their existence as definite bodies marked off in space and separate from other bodies, no mere formless collection of molecules, here to-day and gone to-morrow, like a liquid or a gas; secondly their power of movement; and thirdly their growth, due to their building up more protoplasm by assimilation than what they destroy in the production of energy. These three are all of importance in understanding the origin of organic individuality. Given cohesion of parts, your primeval organism is marked off from the rest of the world. Even though it may be homogeneous, no true system of diverse parts, yet this mere fact of existence as a single and separate material body is a first step towards Bergson's "closed system." In non-conscious animals, indeed, where individuality is bound down within the limits of physical substance, this separateness in space is the only foundation upon which such a closed system could be built. Besides this, it presents itself as a whole unit to the forces of the outer world: living substance thus starts with its foot upon the ladder leading to independence, for its molecules cohere, and all know that union is strength.

Given the complex molecules of fixed composition, and, if of various kinds, existing in a fixed proportion, there will be definiteness of shape and action; and given assimilation—the reproduction of new molecules identical with the old—there is the possibility of continuance for this shape and this action.

Analysed thus far, our organism has revealed itself as very similar to a crystal in its definite boundaries, definite and permanent form, and, we may add, in its capacity for growth. It differs only in having a mode of working as well as a form which is continuous.

The organism, however, has two further properties which make it at once more definite and more independent than the crystal. It is more independent, more self-determining, because it can build up its complicated molecules out of simple substances, and because these substances—its food—may be varied to a considerable extent and the end-result, its protoplasm, yet be the same. A crystal on the other hand, cannot build up the complex from the simple—it can only add ready-formed molecules to its substance, and can only use them if they are presented to it in one particular way, in the condition of a saturated solution.

An organism is more definite because its size is defined as well as its form. A crystal will continue to grow without limit if only the appropriate mother-liquor in which it hangs is kept saturated: its form is definite, its size indefinite. It is a fact of common observation, however, that each organism has a typical size—not invariable, but fixed within certain not very wide limits. This again is due to the differences in the modes of assimilation of crystal and living thing. In the crystal, growth takes place entirely at the surface. Its assimilation is purely physical: it assimilates to its own physical state molecules of the same chemical composition but in a physical state different from its own. Capturing molecules from their state of solution, it builds them up on its solid self in such wise that they fit on to the pattern of the already existing structure.

With protoplasm, however, assimilation is chemical as well as physical, and growth takes place by *intussusception*, not by accretion. That is to say, it works with raw materials,[5] and these materials, instead of being plastered on to the outside, can and do pass in to the interior, and only there are worked up into those combinations of brick and architect, the molecules of protoplasm. Thus, though the absorption of raw materials must of necessity take place at the surface, the actual formation of new living matter, or in other words assimilation and growth, goes on only in the interior.

As a further result of its partially fluid nature, protoplasm is subject to the laws of surface-tension, and a mass of it will therefore tend to become spherical. But in a sphere, as in any other solid body of fixed shape, surface increases with the square, bulk with the cube of the diameter. When we say that one ball is three times as big as another, we usually mean that its diameter is three times as long, forgetting, or leaving implied, that in surface it is nine times, in cubic content twenty-seven times as big. With our balls of living substance, this disproportion between increase of bulk and increase of surface brings difficulties.

Every molecule in the inner parts of the sphere must have oxygen and food if the whole is to go on living. As the organism grows, that is to say as its molecule-population increases, the demand of each molecule is no less, but, owing to the disproportion between surface and volume, the supply available for each is dwindling. The actual materials of supply still exist in unlimited quantity, but the organism cannot get at them. If the English Nation, with population advancing by leaps and bounds, were not able to build harbours and provide dock-labourers as quick as she bred men, all the wheat in Canada, with Imperial Preference to help, would not keep her from starvation, for the simple reason that it could not get in.

So the primeval drop of protoplasm, earliest ancestor of all living organisms, the English Nation not forgotten, found, as it grew, its ports and landing facilities not keeping place with the demands upon them. Each particle of food and oxygen has to be handled by the surface molecules—unloaded from the circumambient water, loaded up again into solution in the general protoplasm—before the central populace can feed or work; and for each four-fold increase of the transport workers there is a sixteen-fold increase within of the mouths to be fed. This cannot go on indefinitely: but what is to be done?

There are two alternatives. One is for the mass of protoplasm to continue its growth, but obviate the difficulty by spreading itself out in one plane. In such a film of uniform thickness, whatever its extent, surface and volume will increase in almost equal proportion. This method, though it has been used here and there, is not easy of adoption, nor wholly satisfactory when adopted. To obtain a thin film instead of an approximately spherical mass of protoplasm, the surface-tension must be very materially altered, and this implies a deep and continuous change in the condition of the surface layer as the size of the whole increases. For main result, the method has the suppression or at least the delaying of reproduction. Logically it leads to unlimited growth of the single mass of living matter, so putting all the eggs of the species in one basket; and even though it is certain that in such a flimsy unco-ordinated film parts would at length be accidentally torn off or simply pull apart from the main body, so reproducing and dispersing the species, yet this reproduction would be long delayed, and the change of structure which involved the delay would have brought few compensating advantages.

The other method is probably easier of adoption, certainly more beneficial in immediate result. It consists in this, that the disproportioned mass of protoplasm divides into two halves. By this means, though the total volume of living substance is left unaltered, the total surface it exposes is increased by over 50 per cent, and the two halves can thus go on gaily growing until the time comes to repeat the process. The actual division seems to be effected by a mere temporary lessening of surface-tension in certain regions, so that this would probably be the way of least resistance for the organism, the way that involved less deep-seated change than the first method. In its results it is certainly better. The species (by which is meant simply the kind of protoplasm), by the repeated formation and subsequent wandering away of new separate masses of proto-plasm, is widely dispersed, so that it no longer presents a single neck by the severing of which some Nero of an accident could with one stroke exterminate the race.

It is thus almost entirely a direct result of the essential properties of protoplasm, scarcely at all an adaptation to outer conditions, that the earliest forms of life defined and limited their size;—in other words, that the first stable phase reached by life in her development on this earth was one in which she manifested herself as a succession of separate protoplas-mic units, each formed from the bipartition of a former one, each beginning its existence as a rounded body of definite but always microscopic size, and each gradually growing, while preserving its form, till its volume was about doubled, when it divided and left its two halves to repeat the cycle. These, the primary units of life are usually called by the name of *cells*,[6] and the cell is the historical basis of organic individuality.

Protoplasm at its first appearance was presumably a homo-geneous substance; as long as it remained so, these masses into which it segregated, however definite their size, their shape, and their reproduction, were yet not individuals. In their working they are like a host of other chemical substances, blindly forging ahead with their reactions, ceasing if the outer conditions transgress certain limits, continuing the same as long as they remain within those limits. They are not, in the strict biological sense of the word, *adapted* to their surround-ings,[7] they are not adapted any more than such a cyclical or catalytic reaction as that which takes place in the manufacture of sulphuric acid from sulphur dioxide, water, and oxides of nitrogen. These, much after the fashion of protoplasm as it builds itself up and breaks itself down, the unstable inter-mediate substance, nitrosylsulphonic acid, must continually make, unmake, and remake itself. As long as the raw materials are present in the right proportions, the reaction will go on indefinitely. So it is with protoplasm: the conditions under which the inorganic reaction can take place are merely more restricted, so that for it to continue, man must step in with elaborate mechanisms to ensure adequate supplies of the sub-stances concerned, provision for their due mixing, care for the removal of their by-products.

Any machinery that protoplasm makes for facilitating its reactions it must not only make itself, but actually out of itself. So it comes about that any improvement in working must mean some change in the structure of the protoplasm, and since improvement usually means division of labour, improved working brings with it a visible differentiation of parts in the previously homogeneous cell. What was a cell and nothing more is now a cell and an individual to boot.

Our primitive homogeneous masses of protoplasm, though all the evidence leads us to assume them, are purely hypothet-ical. Every cell that we know to-day contains at least three, and probably more, diverse and mutually helpful substances. There is the outer layer, whose primary function is absorp-tion, though the secondary one of protection is often added. Its surface-tension and its solubility must be such that bodies which adhere to it and dissolve in it are useful to the whole cell as food. Within this outer sheath are two further substances. One, called chromatin on account of its affinity for many dyes, is chiefly concerned with assimilation, with the constructive part of the protoplasm's chemical cycle. Usually it is all massed to form (together with other substances) a definite body or *nucleus*, but in various primitive forms, such as some bacteria and some flagellates, it appears in the shape of minute granules scattered at random in the mass of the third substance, which constituting the bulk of the cell, is called the cytoplasm. This has as its special duty the destructive part of metabolism; it liberates energy, and uses that energy in doing work, such as locomotion, for the good of the cell as a whole.

All forms of life now living must have had an ancestor which existed under this double form of a cell and an individual, and it is our business now to trace the main lines of this development. Here is no necessity to enter into the causes of change; whether we believe in Natural Selection or Lamarckism, are driven back to Bergson's *élan vital*, or even to a complete confession of ignorance, is immaterial as long as we accept change as a fact. Then our task is merely to trace the change itself in its course and expose what to the best of our belief are the main steps it has taken.

Every organism has a general scheme of architecture which can be seen behind the mass of minor adaptive details. It is easy for instance to recognize the vertebrate plan in such different-looking creatures as a giraffe, a sparrow, and a sunfish, or the insect plan in butterflies and fleas. With such a plan to start from, change may work in three main ways. First, it may run through the variations on the original plan, without introducing any new complication. The different species of a genus, for instance, usually differ from each other in this way. Every one can recognize that polar bear and brown bear and grizzly bear are all built on the same bear-plan, though no one can say that one is better, more differentiated, than another. In the second way the original type of plan is retained, but complications

are introduced which imply true differentiation of parts and
division of labour; such parts have never been free and independent, so that the division of labour is very different, in origin especially, from that of insect communities or our human
society, where the parts themselves begin as independent individuals. This is not mere change for change's sake, but change
progressive. We may call this method *internal differentiation*,
implying that all has taken place within the original unit. An
architectural metaphor may help us. Life finds in the cell the
ground-plan for her first mansion—a one-roomed hut. You
may change your one-roomed plan from round to square,
from square to oblong, and you will not have improved it: but
add a chimney and windows, and at once, though still but one
room, it is something better. Even a church with its aisles and
nave, transepts and choir has grown thus by internal differentiation. In essence it must always be a single room so that the
congregation may see and hear the service; and we realize the
justice with which the Romans used *aedes* in the singular to
mean a temple.

With equal justice they used the plural for a house. They
had reached the stage of civilization when a house was no longer a single room, serving more ends than one at once, and
all in turn, but a collection of rooms, each one different from
any old single-roomed house, all modified in their architecture
from being thus built up into a common whole, but none the
less obviously separate rooms, each in itself a unit, each somehow comparable with the single space of the more primitive
dwelling. This way, of joining unit with unit, is the third way
with organic change. Suppose that instead of separating from
each other after each division, the cells remain connected.
The result will be a colony of cells each one like all its fellows.
If division of labour sets in later among the cells, they are

rendered mutually dependent, and the colony is transformed into a true individual, which is obviously of a higher order than the cell. It has attained what may be termed the second grade of individuality.

This method, for want of a better term, I shall call *aggregate differentiation*, to show that the individual formed by its means consists of an aggregation of smaller individuals. It differs from the second method in that division of labour, instead of taking place among the parts of a single unit, affects whole units or even groups of whole units.

This third method is of special importance for the evolution of life because those organisms that have adopted it have found the only satisfactory solution of that besetting problem—how to become large. It is of importance for the understanding of individuality because it gives the clue to many of the apparent paradoxes of the higher organisms or Metazoa—why they are built up of units comparable with free-living Protozoa, why they so often reproduce by means of a single cell, why the embryo produced from this single cell so often consists of a number of almost identical cells among which division of labour only later sets in.

Now that the animal has separate units to build with, each with a firm membrane and definite shape of its own, progress is much more rapid. In the first place, metabolism can be maintained in spite of increased size, since conducting channels for the food and waste-products can be constructed.

This would be all but impossible within the limits of an enlarged single cell, owing to the semi-fluid nature of its protoplasm; but now since each cell has a firm outer wall, by joining cell to cell tubes can be made through which the food and the waste-products can quickly pass from end to end of the organism instead of having to work gradually through by diffusion. At first, as in flatworms, most of the various systems of the body—the digestive, the genital, and the excretory— are themselves profusely branched, but later the whole business of distribution and collection is taken over by the circulatory system: this alone is ramified and the others can pursue their more proper avocations in peace.[8]

The nervous system is another which can be much more easily perfected in an individual of the second grade. To perform complicated actions which shall be appropriate to the circumstances, there must exist a nervous mechanism consisting of various parts, each part capable of being connected up with every other part. To evolve such a mechanism from a homogeneous mass of substance would no doubt be possible, but to evolve it from a collection of cells would be certainly easier, for there at the outset some of the essentials of the finished product—the separate parts and their discontinuity— would be already given.

Thus through reaching the second grade of individuality, life has been able to gain both size and brain-power for herself. And so it comes to pass that the next steps in her progress have been effected chiefly by the way of internal differentiation. All three ways of change were open to her, and all three have been used in their measure: but the main difficulty, the difficulty of size, has been removed from the path, and the second method can now show its full possibilities. As a matter of fact, the animals of largest size, of greatest intellect, and of best instinctive powers are all individuals of the second grade.[9] At the last, however, when the brain and sense-organs are sufficiently developed, life has gained her most elaborate triumphs of individuality by a return to the third method, of aggregate differentiation. How the method is now modified owing to the possession of a highly-developed brain by the units with which it works, will be treated of in Chap. V: every age has known and wondered at the results it has produced—the communities of bees and ants, and the societies of man himself.

Enough has been said to give the stranger in the land a general orientation, and to show him that Life will guide him to a better view-point than Man alone. The main outcome of the enquiry has been to show that living matter at its first appearance on earth, as the direct results of its material composition, could only express itself in the form of cells—rounded masses, microscopically small, each bound, after attaining a limit of size, to divide into two equal halves. Decreed thus by necessity at the outset, these cells are used ever afterwards as the words out of which all life's poems are fashioned. All living things are made of cells and of structures built by cells: all living action is reducible to cell-action. And it was no hyperbole to say that the English Nation is the direct descendant of an ancestor which throughout its life remained a single cell.

The cell was not from the outset an individual: but by its fixed limits of size, its defined shape, and its power of assimilation (by the combination of these properties, be it understood, and not by any one of them taken singly) it was the first thing evolved to which individuality could adhere. It was like Benjamin Franklin's kite, bringing lightning down from heaven, but it did more than that, for it provided a permanent resting-place on earth where individuality could stay, could gather strength and develop upwards. For this reason it is right to speak of the cell as the foundation of animal individuality. How in later times the relation of the cell to the individual is modified must be left for the present on one side (see pp. 104, 115): we must now retrace our footsteps and see how others have defined the animal individual.

SOME OTHER DEFINITIONS OF ANIMAL
INDIVIDUALITY

From time to time various definitions of individuality have been given by zoologists. Most of them are framed with little reference to the philosophical idea of individuality, and the result has often been that the term individual as defined by them, though applicable to some reality of zoology, can no longer be used without absurdities in its more popular but more correct and more original sense.

One of the most widespread definitions considers the individual as "the total product of a single impregnated ovum" (8 *a*, p. 59), that is to say as the sum of the forms which appear between one sexual act and the next. This would make all the polyps in a colony of hydroids, all the separate polyps budded off by a fresh-water hydra, all the summer generations of the aphis, together constitute but a single individual. Of recent years it has not found so much favour, but Calkins (2) has urged that it should apply to protozoa, declaring that all the separate cells arising by continued division from a single parent between one sexual act (conjugation) and the next, should be considered as one individual, no less than the cells of a metazoan like man, which too arise by continued division from a single parent, the ovum, and remain connected to form his body.

Of the various facts which make the hypothesis untenable, the chief are concerned with the artificial or accidental production of two or more co-existent organisms from a single ovum.

In most animals each single fertilized egg gives rise to a single embryo and this to a single adult organism: but in some, where this is the normal rule, more than one embryo may be accidentally or artificially formed from one egg, and in others this multiplicity is the usual course of events, even though most of their relations may grow up in the ordinary humdrum way—"one egg, one adult."

Aberrations may occur even in man: there can be very little doubt that identical twins[1] (to leave all double monsters out of account) arise from the two cells produced by the first division of a single fertilized ovum, which have accidentally been torn apart instead of staying united.

A very interesting variation on this is seen in the nine-banded armadillo (*Dasypus novem-cinctus*) which regularly produces "identical quadruplets" (14). Most mammals give birth to several young at one time, but usually each grows up from a separate and separately fertilized ovum and each is enclosed in its own set of embryonic membranes. The armadillo's brood, however, like the identical twins in man, has only a single chorionic membrane, and the four resemble each other minutely. Always of the same sex, their measurements are identical; even the number of plates in their armour is constant to less than 1 per cent., though the range of variation from brood to brood may be 5 per cent. and more.[2]

Then comes Experiment and confirms our conclusions of observation. The egg when it develops outside the body of its parent (the rule with most of the lower animals) is at the mercy of the experimenter. After it has divided into two halves, these two blastomeres (as the cells produced by the subdivision of the egg are called) can be separated either mechanically or by chemical means. In the majority of animals where this is possible, the half-blastomere, that identical mass of substance which without man's intervention would have formed half the body of the adult, develops, owing to the mere accident of separation from its sister, into a whole body. Even with such a highly organized creature as the newt this has been accomplished.

The experiment may be carried still further. A whole jelly-fish (Liriope) may grow up from a quarter-blastomere, and in sea-urchins a single one of the first 8, 16, or even of the first 32 blastomeres will make a gallant attempt to develop into a normal whole; and, though it does not succeed, its death seems due to mere minuteness, lack of size, rather than to lack of that internal machinery which produces the complex adult from the simple egg.

These facts are a *reductio ad absurdum* of the theory. It is difficult to consider the two or more experimentally produced sea-urchins or newts as constituting a single individual; the four armadilloes with their one individuality raise more than a doubt; and with the occasional and accidental production of true twins in man comes finality. If anything is an individual on this earth, that surely is man; and yet we are asked to believe that though the most of us are true individuals, yet here and there some man who lives and moves and has his being like the rest is none, that he must make shift to share an individuality with another man simply because the couple happen to be descended from one fertilized egg instead of two. In himself a twin is like any other man; to say that one is an individual while the other is not, takes all meaning from the word.

The idea rests partly on a misapprehension of the sexual process, partly on realities which are of some zoological importance but have no true bearing on the idea of individuality.

Until very recent times the sexual process, the so-called "act of fertilization," was looked on as something which had to be repeated at regular intervals to keep the race going. Somehow it communicated to the organism a mysterious force, which sooner or later dying down must be renewed by repetition of the act.

This is by no means a true view of sexuality. To start with, one large group of organisms, the Bacteria, seem not to possess it at all, while here and there in higher groups it has been lost; the American water-weed (Elodea) for instance, that pest which at one time choked half the waterways of England, started its career in this country by being accidentally imported with American timber, and in all its subsequent development has never been known to form seed.

Lower down, near its first appearance, it is not connected with reproduction at all, as in the Ciliates among the protozoa. Hereto it is not a necessary part of the life-history; Woodruff (20) has recently shown that these animals, which reproduce by fission, may be bred through an indefinite number of generations without conjugation. Enriques, on the other hand (7), has shown that a ciliate which has just conjugated, or in other words received a part of the nucleus of another and joined it with its own, may, before dividing at all, and with this very nucleus just formed by sexual fusion, immediately repeat the process. These observations show that the sexual act stood originally in no relation to the life of the cell, or of the multi-cellular organism, or of the race, so that any conclusions with regard to individuality based on the periodical recurrence of sexual fusion cannot be fundamentally true.

But though the theory cannot be upheld in its entirety, yet some of the facts upon which it is founded are of considerable interest not only generally but also in reference to individuality.

To start with, the upholders of this theory, such as Professor T. H. Huxley (8 *a*), base themselves largely upon the facts of metamorphosis, that sudden change, from the grub to the fly, from the tadpole to the frog, that occurs at a definite point in the life of so many animals. What is perhaps the most remarkable example of metamorphosis, that of the Pilidium into the Nemertine, he does not mention, since it was only established some three years later, but as it illustrates his contentions better than any of his own examples, it may be given here.

Many of the nemertines—salt-water worms with long cord-like bodies—lay eggs each of which develops into a transparent free-swimming creature, very unlike its parent, and called Pilidium from its resemblance to a little hat (Fig. 5). The hat is provided with ear-flaps, and between the flaps there is a mouth leading up into a capacious stomach.

That is its structure when young: at the close of its life, however it is seen to contain a darker something within itself, and this something on closer inspection turns out to be a young nemertine worm, wriggling actively inside a hollow sac which intervenes between it and the tissues of the Pilidium. This is strange; but stranger still, the young worm contains within itself the stomach that was the Pilidium's, so that when the Pilidium feeds, the food passes through its mouth into the stomach which is now the worm's.

The origin of the worm is equally curious: at a certain stage in the growth of the Pilidium, five little pockets appear on its outer surface, arranged in a ring a little above the brim of the hat. The pockets deepen, and their outer openings get narrower and narrower, at length becoming quite "sewn up," so that there are now five closed bags under the skin. These bags flatten and then extend round the stomach of the Pilidium in every direction, laterally as well as up and down; they thus

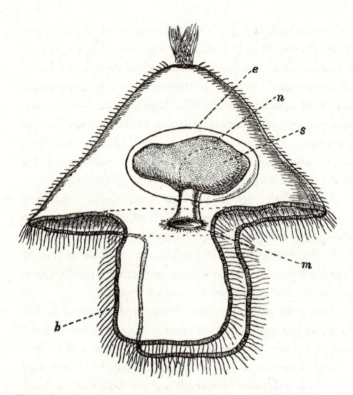

Figure 5
Diagram of a Pilidium with young Nemertine enclosed. *b*, band of
special long cilia; *e*, envelope enclosing the worm; *m*, mouth; *n*, the
young worm; *s*, stomach. (Magnified.)

meet each other, and the walls which are in contact then disappear, so that all their separate cavities join up into one. There
is now beneath the skin an outer shell, then a cavity, and then
an inner shell which surrounds the Pilidium's stomach. This
inner of the two shells or sacs becomes thickened, undergoes
various transformations and at last gives rise to the body wall
and many other organs of the young worm, while the outer
sac is merely a temporary protective envelope. The worm at

length wriggles so violently as to break through this envelope and the skin of the Pilidium, meanwhile tearing the gullet where it passes from the body of the Pilidium into its own. The worm goes on its way rejoicing, and grows up into an adult nemertine; while the Pilidium still swims about, though stomachless, for a time, but perishes at the last.

A perfect gradation in abruptness of metamorphosis can be traced up to this extreme condition. Often, as in man, development proceeds gradually—a slow transition through continual change. In others, as in the frog, there is one period when a sudden alteration of habit and structure takes place; the tadpole, we say, undergoes a metamorphosis and is made a frog. But though there is radical rearrangement, nothing is discarded. In the butterfly there is a more violent metamorphosis, and also a part of the earlier form, its outer skin, is discarded during the change. Finally in the Pilidium not merely the skin but nearly the whole of the larva is rejected at the metamorphosis.

From this, Prof. Huxley then says, it is but one step for the larva to keep all its essential organs when it parted company with the adult form; the one would be formed by the other after the fashion of bud, and from this on to the establishment of colonies like those of the hydroid polyps would be but one step more. Then we should have a perfect transition: the animal as it develops is represented first by a succession of forms, each one turning into the one that comes after, but then first a part and finally a whole of one of the forms comes to have a separate existence in space simultaneously with one of the later forms. So, he argues, since the tadpole and the frog can rightly be called mere forms or phases of the same individual, then the Pilidium and the Nemertine, and then all the polyps in the hydroid colony,[3] are but such forms too.

As a matter of fact, this gradation does not seem to exist in nature; but even if it did it would not be convincing. It is often forgotten that the most perfect quantitative gradation from one condition to another is no guarantee that the two conditions shall not be qualitatively different. To take the simplest example, when the chemical substance denoted by the symbol H_2O is heated, a definite addition to the rate of motion of its molecules is made for each degree of temperature through which it is heated. This quantitative addition, however, has a qualitative result: with continued heating the substance passes from the solid state into the liquid, and from that into the gaseous, turning from ice to water, from water to steam. There is a similar gradual transition in life from the mere aggregate to the higher-grade individual (Chap. IV).

Here, however, there seems to be no such series. All goes well up to the Pilidium, but then comes the gap. There is no case known, where two *complete* individuals are formed as the result of metamorphosis. In reality, the detachment of the Pilidium skin from the young worm is not an attempt at reproduction at all, but is due to something very different. This something is the incompleteness of adaptability in protoplasm, and since the subject will concern us again later (p. 100 *et seqq*.) it may be investigated here.

The whole *raison d'être* of a metamorphosis is the restriction of the animal to one environment in one period of its life, to another and a wholly different environment in another period. Different environments require different structures; and the metamorphosis is the time when the old structures are destroyed. When the tadpole, for instance, suffers a land-change, gills and tail must vanish. They do not, like the skeleton of the gills, become converted, after considerable remodelling, into structures of the adult, nor like the caterpillar's outer skin, are they bodily cast off: they are absorbed, they shrink and their contents are drawn into the body of the young frog for future use, as the yolk-sac and its contents are drawn into the body of the unhatched chick.

The tadpole is so well adapted to the water, the frog so well adapted to the land, that certain organs can not be used, however remodelled, for life on both. They cease to exist as such; it is only the materials of which they are composed, not the living organs themselves, which the animal uses for its further development.

There is a wide possibility of change inherent in all living substance, but after a certain specialization of cell or organ is reached, it becomes impossible to remodel it to perform another totally different function. I say *impossible*: it would perhaps be safer to say that the difficulty of remodelling becomes so great that the simplest way, and so the least wasteful of energy for the organism, is to destroy the old structure, degrading it to the level of mere food-material, and then to build up the new from its very beginnings.

An extraordinary example of this is found in the development of the higher insects. Practically every organ of the body in a larval form like the caterpillar becomes broken down, chiefly by the action of phagocytes, into lumps and masses of dead proteid substances.

A boy known by repute to the writer once expressed surprise that there were any organs inside caterpillars: "I thought," said he, "that they were all just skin and squash." This would be a very accurate description of their condition during the metamorphosis, were it not that embedded in the squash at intervals there lie little patches of living tissue. These so-called *imaginal discs* are formed of unspecialized cells; they grow, unite with each other, and develop gradually into the structures of the perfect insect.

In the Pilidium, it seems, the young worm finds that less energy is wasted in feeding on its own account than in attacking the larval tissues and converting them into readily assimilable food-stuffs. That part of the individual, therefore, which has been so specialized for a free-swimming life as to defy remodelling for worm-purpose, is discarded altogether instead of being absorbed.[4]

As with cells and organs, so with human beings: it is rare that the skilled workman can change his trade. When he is too specialized it may be easier to give him notice and train a new apprentice than to go through the pain and grief of the change from fixed habits.

The remains of the Pilidium then represent merely a part of the Nemertine individual which is discarded as being no longer useful: the history of the process shows that it has nothing to do with ordinary asexual reproduction such as the budding of polyps in a hydroid colony, and so, even were the detached larval part to regenerate a new stomach and become a separate self-supporting organism (as is not unthinkable) we should not be able to draw any conclusions applicable to colonies produced by ordinary fission or budding.

There is another reality on which, unconsciously the theory is based. In all Metazoa there is, before and during the sexual process, a shuffling and recombination of the chromosomes of the nucleus—those bodies which taken together appear to determine the characteristics of the offspring, or at least those which mark it off from others of the same species,—whether it shall be tall or short, fair or dark, chubby or lanky, tip-tilted or Roman-nosed. More, it was supposed that this rearrangement only took place during sexual fusion, and instances were adduced of many vegetable "sports," or mutations as they are now often called, so many of which have been enumerated by De Vries. A plant will often appear showing a mutation in all its parts, so that the change inducing the mutation must certainly have affected the single sexually-produced cell from which the whole plant has sprung. Once formed, mutations will persist in cuttings or slips of the parent plant, but will usually be lost when the sexual chromosome-shuffling is allowed to take place and offspring are raised from seed. In such cases then, all the plants that have arisen thus asexually by grafts or slips, from actively growing parts of the one original parent, all possess, in the mutation, a common character separating them from other plants of the same species, and this common difference persists as long as sexual fusion does not take place between bits of their protoplasm.

Phrases such as "he has a marked individuality," or "he is very individual" lead people to suppose erroneously that one of the chief characters of an individual is its difference from all others. Then, seeking for some clue to guide them through the mazes of animal individuality, they seize upon this and say that because one stream of protoplasm exhibits constant differences from other streams, it is therefore an individual. It then appears that in many cases these differences only persist from one sexual act to the next: therefore, say they, the sum of the forms between two sexual acts must constitute an individual.

However, even apart from the initial flaw, that mere difference constitutes individuality, the chain of argument will not hold, for it is found that not all mutations are similar to those we have described: permanent and considerable changes may take place at any time during the life-cycle, and not in the sexual act alone. The so-called *bud-sports* of many plants are of this nature: from a single bud on a normal tree grows out a shoot displaying some new peculiarity, some mutation which it can transmit to its descendant shoots. A race of trees with the new character can thus be raised by grafting, and not only this, but some bud-sports breed true to seed. Thus nectarines have repeatedly arisen from peaches, not only from peach-seed, but also from peach-buds, and in both cases may subsequently grow true to seed (4, p. 360).

One last partial justification of the theory is left: often when more than a single individual life (in our sense) intervenes between one sexual act and the next, it happens that these several individuals are different from each other but appear in a regular cycle, as in the liver-fluke (p. 16). When this is so, the forms that intervene between two sexual acts do in point of fact together constitute an individuality, one of the type that we have called species-individualities. But this coincidence of sexual act and beginning of a new individuality is only an accident, philosophically speaking, as our previous discussion of the sexual process will easily prove (p. 54).

Thus, though we may note as an interesting fact that the sexual process has at various times and in various ways become connected with one or another form of individuality, yet we must recognize that this connection is not obligatory, that in origin the two are entirely distinct, and that therefore the one cannot possibly be used as the basis for the definition of the other.

Another and a very different view is taken by Le Dantec (11), who, sticking to etymology, gives the following definition:* "l'individu vivant est donc un corps qui ne peut être

---

* "The living individual is thus a body that cannot be divided without at least one of the resulting parts of the division dying."

divisé sans que l'une au moins des parties résultant de la division perde la vie." This happens, he says, only when there exists a nervous system, and one where the nervous elements are concentrated at certain points to form centres of control and co-ordination, a process which as its climax produces the brains of the higher insects and mammals.[5] Each nervous centre constitutes then in some way the nucleus of an individuality and only animals with highly-centralized nervous systems can properly be called individuals.

The real error of this view lies far back in its premises. The definition contains an error of logic. You may correctly insist on etymology and say that an individual is something which cannot be divided without losing its essential quality: but when you say that the essential quality is life, you are not talking sense. The essential quality of an individual is not life but *individuality*. As a matter of fact, an individual as defined in this book cannot be cut in two without its individuality being either lost or impaired (p. 36); and though the loss may be only temporary it is none the less real.

Le Dantec's idea, however, is not merely based on error. The centralized nervous system does form the nucleus, not of any individuality it is true, but of that special kind of individuality, a personality.

However, since not *all* brains, but only those whose mechanism allows some conscious reason and memory, are the structural tokens of a personality, and since it is beyond our present power to discriminate between conscious and non-conscious brains from mere appearance, this structural criterion breaks down in practice and we are driven to accept behaviour as the only accessible touchstone for personality.

The same is true of individuality. An individual is not an individual because it arises from the sexual fusion of two cells, nor yet because it possesses a certain aggregate of white fibres and grey cells called a nervous centre. Even were it a fact that on this earth these two properties were always associated with individuals, they would still not afford the proper basis for a philosophic definition of an individual. They would be mere accidents of the individual, which would still owe its individuality not to them, but to the particular way in which it *works*.

The essential thing about an organism is its actual working, the way it directs the current of energy by which it is continually traversed, and causes it to act on the external world. The main errors of materialism on the one hand and of teleology on the other have resulted from thinking either of substance and structure alone, the mere tools by which the working is carried on, or only of the apparent purpose for which it seems to exist, and not merely of the working itself. Only on this basis can a definition of individuality be attempted, and it is by neglecting this basis that many have been led to false conclusions.

THE SECOND GRADE OF INDIVIDUALITY AND ITS ATTAINMENT

Question: "What's one and one and one and one and one and one and one and one and one and one?"
Alice through the Looking-Glass.

Answer (sometimes):
"Each one almost a Whole, yet all but Parts
They have lost self to form a Greater Whole
Far nobler than its sum of single Parts."
The Green Bayswater.

In Chapter II it was shown that the very existence of the first living individual, the cell, was originally determined by the material properties of living substance. There are large cells and small cells, but, with few exceptions (see pp. 67–68), it is a very limited largeness to which even the largest attain. This limitation, depending as it does upon the surface-volume ratio, is one of the primitive, original attributes of the cell; and to attain size, the cell must in a way do violence to its nature, somehow modifying its surface-tension, overcoming its natural tendency to the spherical, so as to keep its absorptive organ, the surface layer, large enough to supply the demands of the inner mass.

Why, however, should the cell not be content to stay small—what is it to gain from size that it should strive after it? One is apt to think of size as a rather unimportant element in life. With the example of the field-mouse and the elephant, both built so closely on the same type, the wren and the albatross, one comes to think of a model of organization which can be fitted at will on to whatever bulk of living matter is desired. Within wide limits this is true, no doubt, but limits none the less there are.

Many Neo-Darwinians, too, argue that adaptation is the great reality gained by organisms through natural selection, and that, therefore, no one species now alive has preference over any other—for to be alive both must be adapted to their surroundings. But to exist and nothing more, to vegetate merely, is not the fate of all organisms. There *is* a higher and a lower, for some are more independent, more powerful than others.

It is now that the importance of size is seen; for increase of size means increase of independence. Most of the forces of the outside world act only on the surface of the organism; but its own forces spring from the whole mass of its substance. The energy necessary for action is let loose by the chemical breaking-down of the molecules of protoplasm and by their combination with oxygen. This, in a primitive cell, is a function of all the molecules, and so of its total bulk. In the higher animals, where locomotive power is delegated to the muscles, the relation still holds good; the three dimensions and so the shape of the thigh-muscles of a jerboa and a kangaroo are approximately the same, and so the surface-volume ratio will hold accurately. If the *length* of the kangaroo is ten times that of the jerboa, then the *surface* of his thigh-muscles will be a hundred, the *bulk* a thousand times as great. Of the outside forces (all antagonistic or at best passively resistant to the organism) that of gravity only is proportional to its mass. If it alone were to be considered, size *would* make no difference to the animal's movements—the weight to be moved would increase proportionately with the forces that were to move it, since both are proportional to the mass of the whole.

In reality, even if we consider locomotion alone, the resistance of the medium—air, water, or earth—in which the creature moves, is equally important with gravity. Everyone knows how much harder it is for a thin, loose-built man than for a close-knit, compact one of equal weight, to make headway in a gale of wind. That is because the pressure of the wind is proportional to the surface exposed, and the thin man, with relatively more surface exposed, has less muscle with which to drive his body onward.

The home of all primitive life was the water; and the resistance of water is immensely greater than that of air. The disproportion between inner and outer force is here so great that it is as impossible to think of any single-celled animal swimming against the most sluggish river as it is to imagine a butterfly poised steady in a twenty-knot gale.

Once more we see the importance of the surface-volume ratio: but what it preaches now is for the organism in direct contradiction to its earlier lesson. Then it said, "thus far and no further, on peril of starvation." Now it warns, "stay thus small, and be condemned to continue the sport of the elements."

How is life to escape from this quandary? She may be content to remain small, like all the present-day Protozoa. Many of these have attained the most amazing complexity for their size; but there are physical limits to the amount of structures that can be contained in a small fraction of a cubic millimetre of substance, and this way has led up a blind alley. One tribe of plants, the Siphoneae, has made a brave attempt to gain size while still remaining a single cell.[1] A plant of Caulerpa, for instance (one of these sea-weeds), may have several square feet of surface, and in spite of being one continuous piece of protoplasm with a wall of cellulose round it, is differentiated into organs resembling in appearance and no doubt in function the stem, leaves, and roots of higher plants. Needless to say, it has only been able to attain this relatively huge bulk by restricting itself to growth in two dimensions only, and is quite thin and plate-like throughout.

What possibilities of development lay along this line we cannot say; all we know is that actually it has not led far. The real advance has been made in a quite different way; by keeping the cell's original form and plan, but joining up a number of them together so that each preserves a considerable measure of independence, and is yet subordinated to the good of the whole. This resulted in the metazoan type of structure, where the individual is built up out of a number of cells instead of one.

As an example of a simple metazoan, we had better take a primitive sponge. Among sponges, *Clathrina blanca* is one of the most primitive. A graceful vase-like creature, pure white, with a long stalk of attachment, and a mere fraction of an inch in length, it obtains its food, like the majority of sedentary aquatic animals, by producing a current. A stream of water can easily be demonstrated passing out of the circular mouth of the vase, and, with a little more trouble, can be seen to enter by a number of quite small holes scattered over the walls. The current is produced by the cells lining the central cavity (Fig. 6): these stand side by side like sacks in a granary, their free upper ends tapering very slightly, and then truncated at the top. The flat top of each is surrounded by the most remarkable transparent cylinder, a mere film of protoplasm, yet beautifully round, and capable of being drawn in at will, or protruded till it equals the cell in length. This is called the

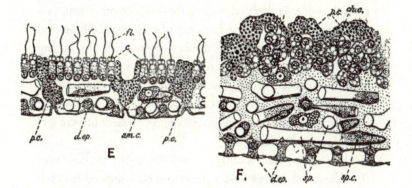

Figure 6
Clathrina coriacea, Mont. Two sections of the body-wall. *E*, not quite fully expanded; the collar-cells line the cavity of the sponge, and show collar and flagellum. *F*, very much contracted. The collar-cells have withdrawn collar and flagellum, and are lying in irregular masses behind the layer of immigrated pore-cells. *am.c*, amoebocytes; *c*, collars of choanocytes (*ch.c*); *d.ep*, dermal epithelium; *fl*, flagella; *p.c*, pore-cells; *sp.c*, spicule-cells. (Highly magnified.)   (From Minchin.)

collar, and in the centre of it there springs from the cell a long vibrating lash or *flagellum*, of uniform thickness throughout, and also capable of retraction within the body of the cell. Usually however it is very much in evidence, beating several times a second, and so producing the current, from which food is taken up and digested by the collar-cells. The bases of these cells rest upon a thin layer of jelly—dead stuff secreted by the living cells, and serving, like the somewhat similar gelatinous tissue we shall see in Volvox, for the common support of the separate cells. On the outside of the jelly is the *dermal* layer of flat polygonal cells, fitting together like a mosaic of tiles. The pores through which the current enters are perforations in the bodies of cells, of a third kind large and contractile, each of which stretches drainpipe-wise from the outer world to the

central cavity. Embedded in the jelly itself are other supporting structures—three-rayed spicules of carbonate of lime, and
through it wander at will a number of amoeboid cells, having
much the same appearance and functions as our own white
blood-corpuscles, except that from their ranks are recruited
the germ-cells, male and female; here, therefore, we have the
unusual[2] spectacle of the germ-cells being pressed into the service of the individual.

Here is obviously a unity, an individual of a higher order
than the cell. Its forms and its functions both depend as much
upon the way the component cells are arranged as upon their
structure; from an examination of a single one of its cells, or
even one of every kind of cell, you could deduce very little
about the properties of the ordered whole. That whole is
greater than the sum of its parts; for the problem is one of combination, not of mere addition. In spite of this the cells have
preserved a very large amount of independence, and indeed
do most forcibly present themselves to the mind as bands of
beings like ourselves that have voluntarily enlisted under some
beneficent tyrant of a general. That analogy, between cells and
men, body and state, has been too often and too far pressed;
its incompleteness is at once grasped with the realization
that no such general does or can exist for the cell-battalions
to obey.

What the bond is that keeps them together, what the
force that orders them—this is still one of the most mysterious problems of life. We must first grasp the extent to which
minor individualities can persist within the major—see how
that centralized empire, the body of one of the higher animals,
was in its origin a federation, not a tyranny.

In Clathrina, the cells' independence is largely realized by
mere inspection. The collar-cells only touch each other with

the lower part of their bodies, and when the sponge contracts, as it does in unfavourable conditions, they—after drawing in their collars and flagella out of harm's way—are actually forced over each other, so that instead of a single unbroken layer there is an irregular collection of cells filling up almost the whole of the central cavity. Whether when the sponge expands again they always fit themselves in between their former neighbours cannot well be proved or disproved, but seems at least unlikely.

The amoeboid cells wander as they please, and the outer or dermal cells, though to be of use to the sponge as protective and contractile tissue they must constitute a single continuous sheet, and so seem merged and lost in the one dermal layer they form, yet show themselves still independent in performing their further function, the secretion of the calcareous spicules. As these are required, single dermal cells break loose from association with their fellows, wander off into the gelatinous ground-substance, and there take up position where the new skeleton is required. Thus, though what they do only has meaning in regard to the whole, the way they do it proclaims them as partially independent beings.

Experiment reveals further lengths of independence— shows the cells capable of veritable insubordination. By means of experiment it has been possible to study the behaviour of the unit parts after the individuality of the whole has been totally destroyed. By chopping the sponge up small, wrapping the bits in the finest silk gauze, and squeezing them, the cells are wrenched from their attachments, and pass through the meshes either singly or at most by twos and threes.[3] By varying the method, one can procure, instead of a mixture of all the sorts of cells, a quantity of collar-cells free from all the rest, and it is their behaviour that concerns us now.

No properly conducted cell, one would have thought, could wish to survive this forcible severance from the whole, the body which we are accustomed to think of as constituting the basis of the only real individuality in an animal. These cells, however, are scarcely inconvenienced. After a short period of shock during which collar and flagellum are withdrawn, they begin joining up one with another, forming irregular solid lumps which, gradually hollowing their central parts, are soon transformed into hollow perfect spheres, their walls a single sheet of cells, and the flagella, now active, beating on the outside. The general resemblance to Volvox (p. 79) is striking, and is made more remarkable by the existence of group of Protozoa—the collared flagellates or Choano-flagellata—whose essential structure is identical with that of the collar-cells; if one of these artificially-produced spheres were found in nature, it would certainly be taken for a colony of Choano-flagellates.

Many of these spheres were kept alive for over a month, and there is little doubt that if the right food were found, they could exist indefinitely, though what would happen with the multiplication of the cells and the consequent growth of the spheres it is hard to prophesy. This remains to be tried; but the facts as they stand are interesting enough. For untold generations no collared cells of a sponge have ever existed except as a subordinate part of a whole sponge-body; and yet, if artificially freed from that "harmonious constellation," they can act independently, can unite into new societies unlike anything known to exist in free nature, and can there subsist for no inconsiderable time.

So much for the independence of the cells: now for their subordination. If, in the experiment narrated above, all the kinds of cells are allowed to remain mixed after their mutual attachments have been broken, we get a result very different from that obtained with the pure collar-cells. First of all, the cells, many of which are still actively amoeboid, and can be seen crawling over the bottom, unite with each other into small lumps and balls. These balls are unlike any organisms known to exist: for, although all their constituent parts are alive, they are without any arrangement and cannot execute any concerted function, Now comes the strange part: this higgledy-piggledy of cells joined up at random is able to reorganize itself, to produce order out of chaos. First of all the collar-cells sort themselves out and form a central solid mass, the dermal cells migrate to the exterior and join up into a single dermal layer. By so doing (though they still resemble no known organism), they have laid down the ground-plan of the sponge, for it is of the essence of sponges to consist of these two layers in this position. The subsequent changes are changes of detail; cells of the outer layer detach themselves and form the spicules between the two layer. Then the inner mass hollows itself out, and the collar-cells (till now quiescent, with collar and flagellum withdrawn) arrange themselves in a single layer round the cavity, and become active once more. Finally an osculum and pores are developed and the random collection of cells (though by processes not seen in normal development) has become an actual sponge, living and functioning, similar in every way to one that has grown up from the egg.

Of the two experiments, the first is the more surprising, the second the more mysterious. In the first, a new form of life is produced—something capable of living, that is, and yet in its structure unlike any known animal: but, given the large degree of independence possessed by the cells, the rest follows naturally. In the second, however, there seems to be a strange organizing power superior in kind to the powers of the cells themselves—an idea of the whole, informing the parts. Again the image of a general directing his army, even of an architect arranging his materials, springs to the mind: but again, where is the general, where the architect? There is no possibility of anything thus extraneous existing in the normal sponge, still less in the little balls, composed as they are of random cells in random grouping. However, the nature of this directive power we must leave for later consideration (pp. 111–112). Here it suffices to have shown that it exists.

So far the analysis of the simple sponge individual has shown it to be composed of definite, separate cells. These in the normal animal have considerable freedom and independence, both structurally and functionally. Under the artificial conditions of experiment, this independence is shown to be very large, inasmuch as one kind of cell at least can live alone, leading a strange new life, when separated from the rest of the body. Though the whole sponge is a true individual, composed of harmonious parts, yet those parts can themselves behave as harmonious wholes. So far, their independence is merely stated and proved; by their history it can be more or less explained, for various converging testimonies all point to one conclusion, that Sponges are descended from a particular group of Protozoa, and that therefore every cell now forming part of a sponge's body is derived by an unbroken chain of cell-division (interspersed of course throughout with sexual cell-fusion) from cells which existed as free-living and independent individuals.

On the other hand there does exist a sponge-individuality higher than that of the cells: to start with, in the normal sponge all these cells are working together for a common end, so that every part helps every other part; and in the second place, the plan of this higher individuality somehow permeates all the cells, so that from any group of all the kinds of them taken at random a whole new individual will organize itself.

After this examination of such a compound individual, we must now turn and trace the method by which this second grade of individuality has been built up, the method by which the Metazoa have evolved from Protozoa. The step from first to second grade is one of the two or three most important in the whole history of life; yet it has taken place successfully on several different lines, and unsuccessful attempts are many.

Among the Protozoa, as among almost all other groups of animals, many species live in colonies—using the word colony to mean a collection of organisms all similar to each other, and all united either by living substance or by some framework that the living substance has secreted.

Such colonies are not higher individuals in any sense of the word, but it cannot be denied that they already possess certain properties on which the higher individuality can be grounded. A colony, besides possessing a characteristic shape, forms a single whole, separate from all other similar wholes; this separateness, as has been seen, is a necessary basis for the exclusively or almost exclusively physical individuality of the lower organisms. As regards function, however, the members of the colony often retain as perfect an independence as they would have if living solitary. Colonial life in such species (which are always sedentary), appears to be merely a device for marking the fullest use of a place with good food-supply. Such spots are few and far between, and are discovered by rare individuals only; thus it is of advantage to retain the descendants of these favoured few bound together there in colonies rather than send them off at once into the world with more chances of failure than of success.

In other colonies, function is not so diffuse, and there is
a function of the whole which is more than, and sometimes
quite different from, the sum of the separate functions of the
parts. Even in sedentary species this can sometimes be seen;
in Zoothamnium, a colonial bell-animalcule, for instance, a
touch on a single one of the animals composing the colony
causes the whole colony to retract out of harm's way. This
general contraction, common to a number of individuals,
though by no means a necessary result of colonial life, could
obviously not occur if the individuals were living separately,
however closely they were crowded side by side. But it is in
free-swimming colonies that the unity of common function
is most pronounced. To take the simplest possible example,
imagine two actively-swimming protozoa of the same species joined together by whatever means you please. If free,
each would have a similar motion to the other, but both
would be independent. When they are joined, however, the
motion of the couple is no longer similar to the motion of
its two components. Mathematically it is the resultant of
their two motions, and as such depends on the way in which
the two individuals are attached to each other. If the action
of their locomotor organs is not fixed and invariable, it will
also depend on the way in which these are used by the two
individuals.

Hence for the couple to move, it is essential that the motions
of its two parts shall not neutralize each other, but that they
shall be co-ordinated to give a resultant motion useful to the
whole couple.

Then there is the resistance of the water to be considered,
so that before a colonial organism can move effectively its parts
will have to acquire a shape, an arrangement, and a mode of

action, differing from those which had served them perfectly when they were independent beings.

The further step necessary before the colony can with full right be called an individual is the differentiation of its members so that they perform different functions. As with the primitive homogeneous lump of protoplasm (p. 43) so with the "homogeneous" colony of similar members; both are on the way to acquiring an individuality for themselves, both exhibit features which are the necessary foundations of that individuality, but neither can with justice be said to possess it.

Illustrating these theoretical points, there exist for us, among various other examples, the members of the family Volvocidae,—an old but well-tried object-lesson. These organisms, claimed by botanist and zoologist alike, are members of the Flagellata, unicellular organisms marked off by possessing long whip-lashes or flagella with which they swim. The Volvocidae seem to be a perfectly natural family. They are all free-swimming; they are all colonial, with a framework of transparent jelly common to the colony; they all possess chlorophyll, nourishing themselves after the fashion of plants; and they all have two flagella, a single "eye-spot" and other morphological characters. There can thus be little doubt that they are all descended from a single ancestor who combined these common characters in his person.

The different forms vary very much, however, in the shape and size of the colonies, in the specialization of the sexual elements, and in the degree of individuation of the colonies.

At the base of the series stands Gonium—sixteen precisely similar flagellate cells embedded in firm transparent jelly, joined in definite arrangement to form a flat disc (Fig. 7). The

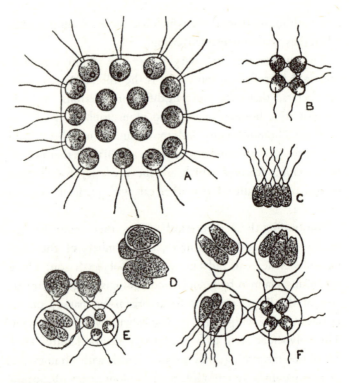

Figure 7
Gonium. *A,* a species containing 16 cells embedded in a flat plate of
gelatinous substance. *B—F,* another species, containing 4 cells. *B, C,*
adult colonies, seen from the top and side respectively. In *D* one, in *E*
two, and in *F* all four cells have divided into four. The four groups of
four cells in *F* will shortly separate and become independent daughter-
colonies. (Highly magnified.) (From West.)

colony thus constituted lives and prospers, nourishes itself,
and grows till comes the time for reproduction. Then each cell
of the sixteen divides—once, twice, thrice, and four times—
into sixteen little ones. Each of the sixteen groups of sixteen
breaks away from the rest, arranges its parts in the familiar
way, and constitutes itself a minute but perfect new colony.[4]

Among all the other members of the family except Volvox, the asexual reproduction (with which alone we need here be concerned) is accomplished in a similar way—each cell takes upon itself to reproduce a whole new colony. They are colonies and nothing more—their members have united together because of certain benefits resulting from mere aggregation, but are not in any way interdependent, so that the wholes are scarcely more than the sum of their parts.

Though, as we have said, Volvox is obviously related to Gonium and the others, it is separated from them by somewhat of a gap.

In the first place, it contains, instead of sixteen or even sixty-four cells, a vast number, mounting up in some species to twenty thousand (see frontispiece). All these cells are inter-connected by fine strands of protoplasm passing through their party-walls[5] and they are arranged in a single layer on the outside of a sphere whose inner parts are filled with a very fluid jelly, so that the Volvox-colony has what we may call an internal medium of its own. Finally, and this is where Volvox has made the great advance, the cells are not all alike. Most are of the type already seen in Gonium and characteristic of the family; these row the colony through the water, steer it, and feed it. Amongst them, in the hinder half of the sphere, are larger cells, lacking flagella and eye-spot, and connected by very numerous strands with their neighbours,

"Their oarsmen-brothers, by whose toil, safe fed
And guarded safe, they live a charmèd life
Within their latticed crystal, peaceably."

And what do they do in return? Now is discovered the skeleton in the flagellated cells' cupboard—they cannot reproduce the colony. They are sterile, and must leave reproduction to the big lazy-seeming cells who are only lazy, however, because they must store up food-materials to start the new colony fairly on its way. They grow and grow, bulge inwards, and finally come to float free in the centre space, where they still grow, meanwhile dividing up into a number of cells. In the end, they become perfect miniature colonies, burst out of their parent and swim happily away.

Volvox is thus a real individual; of the two kinds of cells each has given up something the better to fulfil its own special duty. There is division of labour, and, from the point of view of the species, each kind is meaningless apart from the other.

The division of labour in Volvox is that usually first seen in compound individuals—between the reproductive functions on the one side and all the rest on the other. In other words, one sort of cell is concerned entirely with the species, the other entirely with the separate individuals of which the species consists; to use the current phraseology, the one sort is *germinal*, the other *somatic*. The word somatic opens up another view: Volvox is the first organism which, in the ordinary sense of the word, has a mortal body. Its substance is not passed on unimpaired from individual to individual, but with each act of generation the major part must die, sacrificed for the greater efficiency of the race.

In Volvox, this body consists of but one sort of cell: in all the organisms usually known as Metazoa there are at least two sorts, if not more. Besides the division of labour between germ and soma, there is developed another in the soma itself, at the first between protective and nutritive cells, the one forming an outer covering round the other, which in its turn surrounds an internal cavity. But even if Volvox only possesses species-individuality, the individuality is none the less real; and the fact that in the family Volvocidae we can positively affirm that the step from an aggregate to a higher individual has actually taken place, is one of the most important in biology.

This, however, is not the only way in which the second grade can or has been reached. It is quite possible that division of labour should set in at the very beginning, and that no such thing as a colony, using the word in its usual sense as a number of equivalent individuals all derived from a single parent and still connected together, should ever have existed.

The best examples of animals with such a history are the Catenata, a small group, all parasites of certain marine worms, discovered by Dogiel (6) only four years ago, and containing but one known genus, Haplozoon. The structure of the most primitive member of the group is simplicity itself (Fig. 9, *e*). It is a single row of cells, one end fixed to the wall of the worm's gut, the other sticking out into the gut cavity. The cells, however, are by no means similar among themselves. The first one takes over all the business of attachment, and most of the nutrition. Actively movable, it possesses at its anterior end a piercing spine and a bundle of delicate protoplasmic threads or pseudopodia, which insinuate themselves far up between the cells lining the host's digestive tube, and serve the double purpose of holding the parasite firm and of sucking up the juices of the neighbouring tissue. From its posterior end this head-cell is continually dividing off new cells, which remain attached to each other in series, up to some seven or eight. The hinder cells of the series gradually become filled with particles of reserve food, analogous to the yolk granules in an egg, and finally lose their connection with the rest, dropping off into the digestive cavity and passing thence to the outer world. Attempts to rear them further have not succeeded, but there can be no doubt that their function is reproductive, designed to spread the race to other hosts.

That is the simplest form: thence to the most complex an interesting series may be traced, through species where a few of the hinder cells divide in such a way that the animal's posterior end is a plate, not a mere row, of cells, then up to others where this state of things begins much earlier, so that the plate is broadly wedge-shaped, and finally to forms where the hinder cells divide in all three directions of space, and the posterior end is large and club-shaped, several layers in thickness (Fig. 8). In the front half of the body, little openings exist between cell and cell, which serve to pass food-substances down from the "head" into the other cells. When these are full-fed, they close themselves off from their neighbours and prepare themselves for their reproductive destiny.

The ancestry of these curious creatures is almost certainly to be sought in another group of plant-like unicellular flagellates, the Peridineae. These are two forms which serve to bridge the gap—a large one—between the active free Peridineae and the parasitic multicellular Haplozoon.

The first, *Gymnodinium pulvisculus* (Fig. 9, *a*), is also a parasite but an external one: it is found attached to the skin of various pelagic creatures by a stalk or bundle of sucking pseudopodia like those of Haplozoon. So it thrives till it is full grown: then, breaking off from its stalk, it divides up into a large number of little cells each of which develops two flagella, takes on the form characteristic of the free-swimming Peridineae, and is off to infest new hosts. Here, it will be seen, the same cell devotes itself at one period to nutrition and at another to dispersal.

In *Blastodinium mycetoides* (Fig. 9, *b*), these two functions are carried on by different structural units: the full-fed cell does not break off from the stalk that nourishes it, but divides transversely into two halves which become separated by a

Figure 8
Haplozoon macrostylum, ×300, showing the greatest complexity reached by the Catenata. Only the cell-outlines are drawn. *h*, head-cell with stylet and pseudopodia. A body-cell is being divided off from it posteriorly. (After Dogiel, slightly modified.)

membrane. The one that is no longer attached to the stalk at once begins dividing up to form little flagellates, while the other goes on feeding, grows to full size again, and cuts off a second reproductive cell.

Now imagine the reproductive cells to remain organically connected with the stalk-cell and to be nourished by it for some time after they have been divided off, and you have in essentials a simple species of Haplozoon (Fig. 9, *c–e*).

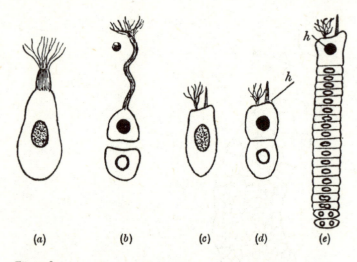

Figure 9
Diagram to show the probable evolution of the Catenata.
(a) *Gymnodinium pulvisculus*, during its nutritive phase.
(b) *Blastodinium mycetoides*. A nutritive cell remains permanently
attached to the host, and repeatedly divides off reproductive cells from
itself.
(c)–(e) *Haplozoon lineare*. h, head-cell.
(c) One-cell stage, resembling (a).
(d) Two-cell stage, resembling (b) except that the two cells adhere to
each other.
(e) Adult, with reproductive cells about to be detached posteriorly.
[Somatic nuclei *black*; germinal nuclei *white*; mixed nuclei *stippled*.]
(Modified from Dogiel.)

The Catenata and Volvox are thus similar in being multicellular organisms with unified function and with division of
labour among their parts; but their origin is very different.

In the making of Volvox, community-life—mere aggregation—came first, division of labour last. In Haplozoon's history, division of labour existed before any trace of communal
existence, and only later was one cell built up upon another
into an individual of a higher order.

To take parallel cases in a different grade of individuality, the simpler Volvocidae closely resemble many low human races among which every family exercises all the ordinary arts and crafts and where society, in spite often of strong communal life, can therefore not rise above the dead level conditioned by the impossibility of doing all things at once and doing them well. Curious and interesting it is that these same peoples if taught, can generally learn and learn quickly and well, many arts and industries before undreamt of among them. The capability was there, but they had not learnt how to use it: only by sacrificing some of their multifarious functions is it humanly possible to advance in the rest, and so to raise society. As with men, so with cells—a jack-of-all-trades cannot advance in any, and the same lesson of sacrifice has to be learnt before the colony can become an individual organism.

A human illustration for the methods of Haplozoon may also be found, or at least imagined. Imagine then a man inflamed with the desire to spread among a benighted race some gospel of good tidings. Poor, he prints the books himself; then comes the question of sending them forth. It is obviously impossible for a single man to do one and the other simultaneously. If he goes out to distribute them himself, the printing will be at a standstill while he is away. If, however, he can obtain volunteers to distribute the books, he himself can stay behind and pull off impressions all the time while a new man goes off with each consignment. Suppose further that while printing he can instruct the distributors in such a way that they will later be able to do their work more soundly, then there will be collected a crowd of embryo distributors at headquarters, from which the fully-trained ones will from time to time depart.

In the first stage the business is like *Gymnodinium pulvisculus*: then like Blastodinium, and at the last like Haplozoon:

the division of labour has come as the first forerunner of the higher development, and this it has done because in both cases there are two special functions to perform which cannot be performed simultaneously by a single individual.

The existence of two mutually exclusive necessities is thus the origin of this type of higher individual: at first the single cell performs them both, but at the expense of not feeding while it is reproducing, not reproducing while it is feeding. Again the sacrifice by a part leads to improvement for the whole; the great fact once discovered that of two cells one can feed for both, the other reproduce for both and the later steps follow almost as a matter of course.

It is to be remarked that for the two functions of nutrition and reproduction thus to clash with one another, it must needs be that the organism can only thrive in a very special and a very limited environment. An individuality like that of the Catenata is therefore found chiefly among parasites, which exist in just such an environment; but in the outer world the conditions are rarely narrow and rigorous enough to call forth such adaptations.[6]

It was the other method, aggregation of similar units and subsequent division of labour among them, that opened to life the full resources of the second grade of individuality. In some colony like Volvox there once lay hidden the secret of the body and mind of man.

THE LATER PROGRESS OF INDIVIDUALITY

"It is provided in the essence of things that from any fruition of success, no matter what, shall come forth something to make a greater struggle necessary."
WALT WHITMAN

Every human being who has passed through the moral struggle will testify to the truth of these, Walt Whitman's words, in their own experience: the biologist will witness that they symbolize as real a truth in the history of life. Life can never be in equilibrium. Given the two well-established facts, that living substance can vary, and that living things if left to themselves would multiply in rapid geometrical ratio, then change in the *status quo* is inevitable. A state of equilibrium may for a time exist, but every balanced organism is as it were pressing against every other, and a change in one means a rearrangement of them all.

The correlated evolution of weapons of offence and defence in naval warfare is closely similar, though simpler far. The leaden plum-puddings were not unfairly matched against the wooden walls of Nelson's day. Halfway through the century, when guns had doubled and trebled their projectile capacity, up sprang the "Merrimac" and the "Monitor," secure in their iron breast-plates; and so the duel has gone on, till now, though our guns can hurl a third of a ton of sharp-nosed steel with dynamite entrails for a dozen miles, yet they are confronted with twelve-inch armour of backed and hardened steel, watertight compartments, and targets moving thirty miles an hour. Each advance in attack has brought forth, as if by magic, a corresponding advance in defence.

With life it has been the same: if one species happens to vary in the direction of greater independence, the inter-related equilibrium is upset, and cannot be restored until a number of competing species have either given way to the increased pressure and become extinct, or else have answered pressure with pressure, and kept the first species in its place by themselves too discovering means of adding to their independence. While the balance of power lasts, variation no doubt takes place, but there is no strong necessity to guide it. Once let a large, favourable variation take place in a species, however, so giving it a handicap, and then for its competitors natural selection is at once made more active—they must perish or else adjust themselves by a variation, generally in a similar direction. So it comes to pass that the continuous change which is passing through the organic world appears as a succession of phases of equilibrium, each one on a higher average plane of independence than the one before, and each inevitably calling up and giving place to one still higher.

This digression was necessary to give some explanation of the succession of ever more perfect individualities, and the continual repetition of the same methods in their attainment.

Space forbids more than the merest outline of the developments of individuality after it has attained its second grade. We have seen (p. 48) that the method of aggregate differentiation is now for a time the less important: it still, however, exhibits some interesting points.

To start with, division of labour in colonies of second-grade individuals, as in colonies of cells, almost universally sets in first between the nutritive and the reproductive functions— the somatic and the germinal.

The Hydroids and their relations give us a series closely parallel at first, though in a different grade of individuality, with that of the Volvocidae, but exceeding it considerably in length. Hydra, one of the simplest hydroids known (Fig. 3), has, like all others, the power of budding; but its buds eventually become detached, so that it never forms more than a very small and temporary colony. Besides this, there is no division of labour among different polyps; all are alike, and whether they shall reproduce sexually or asexually is dictated to them by the external conditions.

Then come the colonial forms: and all of these show some division of labour. All, for instance, when ripe, bud off special sexually-reproductive individuals in the shape of little jelly-fish or *medusae*. Sometimes any polyp of the colony may be able to give birth to one of these, but very often the ordinary polyps reserve themselves for feeding, and special mouthless polyps exist for the one purpose of budding off the jelly-fish; they are the producers of the reproducers of the colony. It seems to be only later that the somatic functions, the functions pertaining to the single colony as opposed to the race, get differentiated, as in Hydractinia (Fig. 10) where there are special polyps that defend the colony as well as those that nourish it.[1]

All these species are sedentary for most of their lives; history again repeats itself, for once more it is in the free-swimming forms that the members of the colony have been most modified, most subordinated to a higher individuality.

The Siphonophora, close relations of the hydroids are a group of beautiful pelagic creatures, which slowly drive their

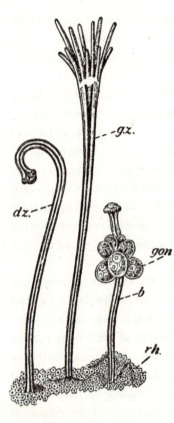

Figure 10
Part of a colony of *Hydractinia*. *dz*, dactylozooid (defensive person); *gz*,
gastrozooid (nutritive person); *b*, blastostyle (asexual reproductive
person); *gon*, gonophores (sexual reproductive persons); *rh*, hydrorhiza
(creeping stolon). (Magnified.) (After Allman.)

trailing length through the water by an array of pulsating bells. Besides these locomotor organs, there are in the colony organs for feeding, for reproduction, for defence, for offence, and for flotation (see Fig. 11). Most of these apparent organs are really modified individuals, either of the polyp or medusa type. The reproductive "organs" are sometimes detached as perfect jelly-fish; the swimming-bells and the protective bracts often show unmistakeable vestiges of medusoid structures. The nutritive "organs" are very like an ordinary polyp, but without tenta-cles, and the defensive organs usually have a structure like that of the defensive polyps in a hydroid like Hydractinia.

There can be no doubt therefore that the various "organs" do really represent modified hydroids; and as these are them-selves individuals of the second grade, a Siphonophoran is therefore an individual of the third grade. The same process—the subordination of the lower individualities to a higher—which we traced from a simple flagellate up to Volvox is traceable again from Hydra to the Siphonophora: but the interesting point is that nowhere else in the animal kingdom is there an unbroken series—a series in which we can affirm positively that beginning and end are close relations—of such length. In the majority of Siphonophora, the persons of the colony have mostly only a historical individuality: some of them are sometimes so much modified and reduced that it has baffled all the zoologists to decide whether they are homolo-gous with individuals or with mere appendages of individuals: and in function each is devoted so little to itself, so wholly to serving some particular need of the whole, that if one were separated from the rest, it would appear a perfectly useless and meaningless body to an investigator who did not know the whole to which it belonged. There are zoologists who say it is incredible that the cells of the Metazoa can be homologous

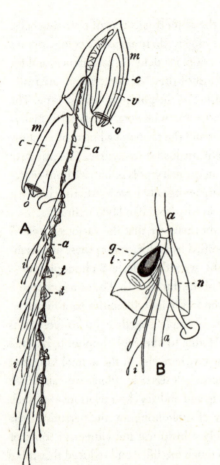

Figure 11
A., *Diphyes campanulata* (natural size). *B.*, a group of appendages
(cormidium) of the same *Diphyes* (magnified). (After C. Gegenbaur.) *a*,
axis of the colony; *m*, nectocalyx (swimming "organ"); *c*, sub-umbral
cavity of nectocalyx; *v*, radial canals of nectocalyx; *o*, orifice of
nectocalyx; *t*, bract (protective "organ"); *n*, siphon (nutritive "organ");
d, gonophore (reproductive "organ"); *i*, tentacle (defensive "organ").

with independent beings like the Protozoa, impossible that a colony should ever give rise to a single individual of a higher order than its members (see 5, p. 304). To them we commend the example of the Siphonophora, and pass on to consider some other individualities, formed through aggregate differentiation, but after an entirely new fashion.

To start with, we have the old but ever interesting fact of *symbiosis*, where two organisms as it were inter-penetrate, entering into a very close relationship from which both parties derive profit. The classical examples of symbiosis are the Lichens, which, long supposed to constitute a distinct group of plants, were in the middle of last century discovered to be actually a mixture of two organisms, one a colourless fungus, and the other a green plant—a simple alga. For the details of their organization any textbook of botany can be consulted: here it must suffice to say that there is a perfectly definite arrangement of the algal and fungal constituents. The interesting thing about them is that they will grow, as anyone can see for himself, in situations which no other plant would tolerate, so that both plants must obviously derive advantage from the combination. Put very briefly, the facts are these: fungi can only get the carbon of their food from organic matter, while green plants have the power of using the energy of light to appropriate carbon from the carbon dioxide of the atmosphere. In respect of the absorption of water and mineral salts, however, the fungus seems to be the better equipped. Thus (division of labour once more) the alga supplies carbon compounds for both, the fungus looks after most of the rest of the nutrition, and also shelters the alga from frost and drought.

Algae identical with the green cells of the lichen are found free-living in spots less exposed than those where lichens grow: the alga, that is to say, *can* exist separately, but in the

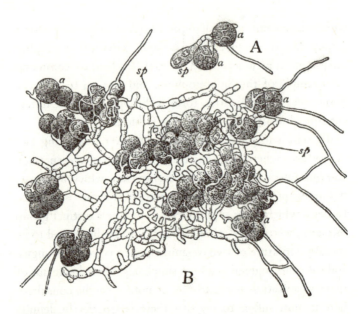

Figure 12
Physcia parietina; building up of the Lichen out of the Alga and the
Fungus. *A*, a germinating fungus-spore (*sp*) which has seized upon two
cells (*a, α*) of the alga *Cystococcus humicola. B*, more advanced stage. The
spores of the fungus have formed a network in which are embedded
numerous algal cells (×400). (From Scott, after Bonnier.)

partnership grows more luxuriantly and has a wider range.
The fungus, on the other hand, though it has been grown separately in an artificial medium, cannot develop in nature unless
it meets with some algal cells. Fig. 12 shows a young fungus
which has just germinated among a group of algae and is now
sending forth little tentacles to seize and enwrap them.

The fungus gains more than the alga, but this does not prevent the combination of both, the lichen, from being a very
definite individual. A lichen on a barren rock is a something
whose continuance as such and in such a situation is dependent upon the co-operation of its two constituents. Here the

division of labour is given beforehand in the two kinds of plants, and the new individuality simply arises from the fitting together of these two separate beings into a very close and special relation.

This relation is, however, only a special case of the general relation existing in nature between green plants and all other organisms. Put very crudely, the relation is this:—green plants can build up protoplasm from water, carbon dioxide, and mineral salts: the protoplasm thus formed is the ultimate source of all nourishment to the rest of life. Animals either eat green plants or also eat other animals that eat green plants; many bacteria feed on the dead tissues of animals and plants, bringing about as a result of their activity the phenomenon known as decay; and fungi live to a great extent on the substances produced during decay. Meanwhile, however, the waste products of the current of metabolism and the final products of decay, which come eventually to be degraded to simple stable substances such as water, carbon dioxide, ammonia, and nitrates, get diffused in the soil, and form the basis once more of the green plant's activity.

In a sense, therefore, the whole organic world constitutes a single great individual, vague and badly co-ordinated it is true, but none the less a continuing whole with inter-dependent parts: if some accident were to remove all the green plants, or all the bacteria, the rest of life would be unable to exist. This individuality, however, is an extremely imperfect one— the internal harmony and the subordination of the parts to the whole is almost infinitely less than in the body of a metazoan, and is thus very wasteful; instead of one part distributing its surplus among the other parts and living peaceably itself on what is left, the transference of food from one unit to another is usually attended with the total or partial destruction of one of the units.

Within this biggest system, nature has been persistent in her efforts to create other "naturally-isolated systems," other individualities. Out of every little accidental company she tends to make an inter-related whole whose parts are largely dependent on each other, and only slightly on other wholes or their parts.

It will be necessary to give a few more examples of the inter-relation of two distinct species before developing this idea. A very instructive example is that of *Convoluta roscoffensis*, a marine flatworm. Its story has been so clearly told by Prof. Keeble (10) that here an outline will be enough. In nature, the worm is always associated with a unicellular green plant which lives in great numbers beneath its skin. The plant on the other hand is found abundantly apart from the worm, but swarms round the egg-capsules in order to procure nitrogenous food, and gets ingested by the young animal. Unless this happens, the worm cannot develop further—the presence of the green cells is the only stimulus which will start its machinery on the next stage of its working. At first both members gain from the association, much as in the lichen, but finally, after the worm (which at last takes no food, but depends entirely on the surplusage of the alga) has produced its eggs, it finds itself short of nitrogenous material, and begins attacking and digesting the green cells; they cannot last for ever, and when they are all gone last of all the worm dies also, trusting to chance that its young will find new algae. This shows a transition from symbiosis to parasitism, though the host here enters the relation of its own free will. Convoluta somewhat resembles an employer of slave-labour in a country where slaves are very kindly treated: the green slaves are well provided for during their individual lives, but they have sacrificed the power of further perpetuating their species. A growing Convoluta plus its contained green cells is therefore that anomaly, a temporary individual.

Take next a case of true parasitism. With most internal parasites, such as trypanosomes (the flagellates which cause sleeping-sickness and other diseases), or tapeworms, each species of parasite is confined normally to one host-species, and cannot come to perfection elsewhere. It is often extraordinarily closely adapted to its environment both in structure and life-history, as a study of any tapeworm will show; but that environment is an extremely limited one.

After this consider an apparently very remote subject—the relation between insects and flowers. I need here merely point out that many insects, such as bees and butterflies, procure all their food from the honey or pollen of flowers, and that most plants with conspicuous flowers rely exclusively or chiefly on insects for their fertilization, and so for their continuance as species. Both insect and flower have been radically modified in structure and appearance through this mutual relation. Most flowers are fairly catholic in their tastes, and are adapted for fertilization by a number of different insects, and the same is true, *mutatis mutandis*, of the insects. But sometimes the relation is a much narrower one, till finally an insect may be able to get food only from one particular flower, the flower to be fertilized only by this particular insect. A relation of this degree of intimacy (though with not quite the same purposes) is found between the Yucca-plant and a moth of the genus Pronuba (Fig. 13). Here (for the details I must again refer the reader to other books; e.g. (17), Vol. I, p. 201) the Yucca can in nature only be fertilized by the one agency of the moth: she, when the time comes for egg-laying, flies to the Yucca, rakes up a large ball of pollen by means of a unique structure on her head, and then flies with the ball to another flower; there she

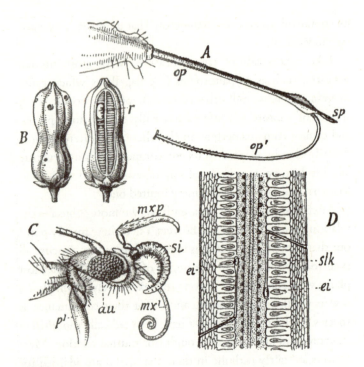

Figure 13
The Yucca and its Moth (*Pronuba yuccasella*). *A*, ovipositor of the moth.
op, its sheath; *sp*, its apex; *op'*, the protruded oviduct. *B*, two ovaries of
the Yucca, showing the holes by which the young moths escape, and (*r*)
a caterpillar in the interior. *C*, head of the female moth, with the
sickled-shaped process (*si*) on the maxillary palps for sweeping off the
pollen and rolling it into a ball. *mx'*, the proboscis; *au*, eye; *p'*, base of
first leg. *D*, longitudinal section through an ovary of the Yucca, soon
after the laying of two eggs (*ei*). *stk*, the canal made by the ovipositor.

sticks her long and curiously-shaped ovipositor into the ovary
of the flower, and lays her egg among the unfertilized seeds
inside. Last she lifts the pollen ball on to a special hollow on
the top of the stigma, and pokes it firmly down. The pollen
fertilizes the ovules, of which there are about two hundred,
and they start developing into seeds; meanwhile the caterpillar

hatches, and feeds at the expense of the seeds. However, it only needs some twenty or so before undergoing its transformations into pupa and moth, and leaves the rest to grow into new Yucca-plants.

The whole proceeding is of great interest, showing as it does the blind and instinctive nature of the organisms' actions, and giving us an example of two species absolutely dependent on each other for their continued existence. If the moth had not the structure to form the pollen-ball and the instinct to put it on the stigma, the ovules would not be fertilized and her offspring would have no food; and if the plant was not prepared to sacrifice some 10 per cent. of its brood, the rest would never develop at all. Here it is of course the two species that are affected, while the single moth and the single plant do not depend on each other in any way; but the essential point of the relation—the mutual helpfulness of two unrelated kinds of protoplasm—remains the same.

Now return and consider these various relationships from the point of view of individuality. The different species of living things and their members are all bound up, though but loosely, into a general whole. Any single species relies on others for some of the necessities of its existence. In many green plants this dependence on other species is very slight and very indirect, while animals, owing to their mode of nutrition, are always directly dependent on the one or many organisms on which they feed. None the less, in nett total of true independence most animals are far ahead of plants. They have had to make more effort to get their food, and throughout life, effort always seems to bring in its train advantages, unforeseen and unconnected with the effort's immediate object. To give an extreme example, the eyes and ears and other sense-organs of animals were developed chiefly for the capture of prey and the avoidance of enemies; but once formed, they were the starting

point for the life of consciousness that has culminated in ourselves. A blind deaf-mute child can be fed and live a healthy
physical life; its mind, however, scarcely exists:

"for the book of knowledge fair
Presented with a Universal blank."

But wisdom at one entrance still can find a way—through the
gateway of touch; and the story of Helen Keller, the American
girl who became blind and deaf and dumb in infancy, will show
how absolutely dependent on external stimulus, even in its deal-
ings with the abstract and the non-spatial, is the mind of man.

The necessity for effort—the "struggle for existence" in the
most general sense—has from age to age raised the average level
of independence, the measure of individuality's perfection in
living beings. In spite of this general rise of level, there has been
in every age a falling away, a decline in perfection of individuality in certain species. This decreased independence reveals
itself not only as structural degeneration, but also in degeneration's opposite, structural specialization. There is, however, a
common cause beneath these opposed effects, and that is over-
close adaptation, adaptation to very narrow conditions.

It is self-evident that all organisms must be more or less
adapted to their surroundings; in other words they must be
more or less dependent upon their environment. Failure to
exist in any but a very limited environment is obviously a
weakness, a lack of independence, and it seems to be a fact
that adaptation to any such limited environment makes it
impossible or very difficult for an animal to exist in any other
environment. The very success of the adaptation decreases the
creature's adaptability.

The adaptation may be concerned only with inorganic
nature, as when plants are adapted to conditions of tempera-

ture, light and moisture, or only with other animals or plants, or with both. Let us take the second as being most germane to our present purpose. The nutrition of animals falls within this province, since they are always dependent on the protoplasm of other living species for their food. This is a limitation, but its boundary is a wide one. The animal may either make an effort[2] to secure its food, or it may prey parasitically on the labours of another animal. Both ways, if too special adaptation is allowed, may lead to a back-sliding in individuality. We can take a series of examples from birds. The Rook and the Sparrow are almost omnivorous, and therefore very independent as regards food-supply. A bird like the Swallow is a little more dependent. In its large gape and strong flight it shows a general adaptation for catching small insects on the wing, but as long as they are insects and small and flying, it is content; it has taken advantage of a common property of many insects, and is dependent in no narrow way. Dependent it is, however, and when the insects fail it must migrate.

Finally, such a bird as the Skimmer (*Rhynchops nigra*) exhibits a very special adaptation indeed. Darwin (3, p. 137) gives a graphic account of them. "The beak is flattened laterally. . . . It is flat and elastic as an ivory paper-cutter, and the lower mandible, differently from every other bird, is an inch and a half longer than the upper." When feeding, "they kept their bills wide open, and the lower mandible half buried in the water. Thus skimming the surface . . . they dexterously manage with their projecting lower mandible to plough up small fish, which are secured by the upper and shorter half of their scissor-like bills." This strange bill is without doubt an extremely efficient instrument for catching fish near the surface of the water, but the length of the lower mandible, the very particularity which makes it so efficient for this one purpose, renders it unavailable for any other. The narrow domain where air and water meet it has made its own, but to that one domain it is rigidly confined.

The path to parasitism, in spite of apparent differences, is very similar. Here too what the animal seeks is adaptation to an environment which by very reason of its peculiarity and narrowness is not already occupied by other competitors. Eventually the fate of the parasite becomes bound up with that of the host. The final result is thus the same; the form which has made the too-special adaptation loses independence.

Such happenings the phrase-monger will find place for under some vague, comprehensive title such as "Filling a vacant place in the Economy of Nature," and be content. But, though it is undoubted that the pressure of the struggle is always forcing life into these vacuums of vacant spaces, we have to look further before we find what the effect on life will be. Then we see that the process is not always so wholly satisfactory as phrase would make it. In our particular cases the result of "filling a vacant space" is that one species gets preyed upon, and the other, the claim-staker, though gaining the gold in the vacant claim, thereby ties himself down to that little plot of ground, and sinks in the scale of individuality.

Now suppose that the one organism does not merely rush into a ready-made vacuum provided by the other, but that the two should conspire together to create a vacuum of their own, into which, as fast as it is created, they jointly creep. This is in effect what happens when two species become mutually dependent. Here again the relation is at first a general one, as between insects and flowers, but at the last may get very special, as between the Yucca and its moth. Both species here have lost independence. The Yucca, for instance, has to be propagated artificially in Europe, for when it was brought over the moth was left behind, and so no seed can be set. At first sight, then, such a system appears like a double parasitism, and twice the evil that parasitism brings should be its portion. This is not really so, for while the true parasite takes what he can get and gives nothing in return, here each pays the other willingly, for services rendered. In extremes of parasitism there is maximum waste; mutual aid (though it implies mutual dependence) establishes minimum waste. Moth and Yucca together consti- tute a system which is harmonious and economical because division of labour is at work: each does what it can do best and gives of its superfluity to its partner. If the two parts have sunk in the scale, yet by that very sinking the beginnings of a new whole have sprung up. They have lost in independence, but something else—the system formed by their combination— has gained in harmony. Put in other words, their own individ- uality has become impaired, but this is compensated for by the formation of a totally new individuality, rudimentary though it be. If their parts of the system, instead of being related by but one tie and for a short space only, were to be brought into relation for the whole of their lives, the resulting system would have the chance of becoming not only more harmonious, but even a more independent individuality than was either of its parts before their mutual adaptation—a consummation actu- ally realized in the Lichens.

This necessity for the parts of a compound individual to lose their own independence for the ultimate greater inde- pendence of the whole—this increasing mutual parasitism of the units within an individual—is in fact a brief statement of the main facts observable concerning internal differentia- tion. Internal differentiation, indeed, to be strictly accurate, is the only way in which individuals are formed, for aggregate differentiation is only a convenient label for the combination of two processes—first the forming of an aggregate, be it of molecules, cells, or persons, and then the welding of this mere aggregate into a true individual by means of internal differentiation.

The progress of the individual of the second grade is in essence a progress towards greater complexity, more harmonious co-ordination, higher independence; this is revealed to the eye in the multiplication and specialization of its various kinds of cells. When it is reflected that these lesser individuals were originally all alike and all self-supporting the modifications which they have undergone are nothing short of amazing, as a glance at any text-book of histology will show. We can but mention a few of the most remarkable. For shape, the ordinary nerve-cell (Fig. 14) is striking enough; the cell-body is an ordinary, somewhat polyhedral mass of protoplasm, but from it are given off branching processes which divide and sub-divide and with their finest sub-divisions come into contact with the branches of other nerve-cells; and at one point runs out a single thread, the nerve-fibre, which, though its thickness is to be measured in hundredths of an inch, may yet reach a length of several feet before it finds the muscle it is to move or the sense-organ whose message it is to carry. Remarkable in another way are the epithelial cells of our skin, which, continually produced in the deeper layers, undergo a gradual metamorphosis into the thin plates of horny matter called scurf, or scarf-skin; by their perpetual wearing off and replacement from below, they give us an outer covering which shall be at once pliant and sensitive, of considerable strength, and quick-healing.

When an individual's only duty is to commit suicide for the good of the society to which he belongs, the process of subordination has gone a very long way. In what are known as *syncytial* tissues it has gone perhaps still further: here the cells, surrendering all separateness of existence, have fused to form mere sheets of continuous protoplasm studded with nuclei. These syncytia are only the final outcome of a process whose beginnings we saw in Volvox and Haplozoon—the connection of the separate cells by means of strands of protoplasm passing from one to another. These connections seem to exist in all multicellular green plants, and have now been demonstrated in a great number of animals.[3] Many zoologists indeed believe that the fine endings of the nerve-fibrils are always in direct protoplasmic continuity with the organ they supply.

It is worth remembering that the actual history of every individual runs roughly parallel with what we know of the history of the race. Those who cannot bring themselves to believe that the ancestor of a nerve-cell, for instance, was once independent and capable of all the functions of separate existence, often do not consider that every nerve-cell started its life simple, rounded, and undifferentiated, only later throwing out its complex branching processes, and later yet coming into protoplasmic continuity with other cells, originally far distant in the body (Fig. 14).

In many animals indeed every individual epitomizes in its own the history of the race. Starting as a fertilized ovum, an individual of the first grade, it next becomes a "colony" of nearly similar cells, and then an obvious second-grade individual, with outer protective and inner nutritive layer. Then comes the internal differentiation of this individual: blood-cells and blood-vessels, nerve-cells and sense-organs, muscles, sinews, bone, kidney, liver—kind after kind of cell is created. And let it never be forgotten that in the embryonic development of any and every individual all these and many others are descendants of a single and a simple cell.

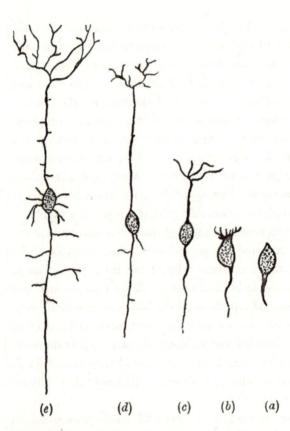

(e)   (d)   (c)   (b)   (a)

Figure 14
Five stages in the early development of a nerve-cell from the brain. In
(a) the rudiment of the nerve-fibre is seen. In (c) the dendrite with its
branching processes has become obvious. (e) does not represent the
final stage; in the adult nerve-cell size and complication are many times
greater. (Highly magnified.) (After Ramon Cajal.)

It is noteworthy that the course of internal differentiation has over and over again—in worms, in insects, in crus-
tacea, in spiders, in molluscs, and in vertebrates—tended
in the same direction—towards the formation of a Brain.
Brain-development has usually gone hand in hand with the

specialization of other organ-systems—the brain seems a mere bit of machinery necessitated by the complexity of the other parts. In the higher insects and the higher mammals, however, the brain seems to have transcended all the other parts of the body, to have gone farther than they in specialization, and to be now in truth the master by whom the rest are to be employed.

This development of sense-organs and brain has had great influence on the progress of individuality. We do not usually stop to consider in what dense darkness the majority of living things must live and move and have their being. Without brain or sense-organs, theirs must be a dim and windowless existence. The world lies waiting round about; but it cannot make its way into their beings. Or say, the world is locked, and living things must make their own keys to it. So it is with men: educating their minds, they forge key after key, each opening a new chamber, letting in a new flood of light from every material object. Show a flower to an aboriginal savage: what he sees is something very different from what Wordsworth or Sir Joseph Hooker would have seen. What he sees, however, contains more of reality than what a beetle or a snail, with their imperfect eyes, could see. The effect produced on an organism when some object is presented to its senses thus depends partly on the perfection of its sense-organs, partly on that of its brain. As we go down the scale, both dwindle: veil upon veil is let down: till at the last there is an almost utter darkness, and not of sight alone.

It is this darkness at the base of the animal kingdom which has there made it almost imperative that the parts of an individual should cohere physically; separate them and they would be lost, and could never enter again into their mutual relationship. Once produce a sense-organ, however, which with the brain behind it is capable of clearly perceiving and accurately localizing distant objects, and it at once becomes possible to construct an individual, such as an ant-community, whose parts, though not contiguous in space, are yet bound together as fast as the cells of a sponge or the persons of a Siphonophoran; here as elsewhere the real bond is an impalpable one—mutual dependence.

The communities of ants and bees are undoubted individuals. Wheeler in a recent paper (18) has abundantly justified this view from a somewhat different standpoint. Here I can only say that if the ideas and definitions put forward in Chap. I are accepted, their individuality is beyond dispute. In spite of space, I cannot refrain from giving one example of the lengths to which internal differentiation of parts can go in such apparently loose-connected wholes. In several species of ants there are special workers whose duty it is to imbibe honey till their fair round bellies are drum-tight, then to suspend themselves, a row of living jars, from the roof, and there to wait until their store is needed by the colony and they are taken down and tapped for general consumption.

One interesting property gained by brains and sense-organs:—organisms possessing them can easily enter into more than one individuality. The Yucca and its moth, for instance, constitute a definite individual that works for its own perpetuation. But their time of contact is a short one; and there is nothing to prevent the moth from entering into relations with some other flower for the sake of food (in return of course fertilizing the flower) and so forming together with it another "whole with inter-dependent parts working for its own continuance"—another individual.

When we come to man, this power possessed by one unit of entering into more than one individual "at once" (see p. 127n4) becomes very marked. A man can very well be at one time a member of a family, a race, a club, a nation, a literary society, a church, and an empire. "Yes, but surely *these* are not individuals,"—I seem to hear my readers' universal murmur. That is a question which neither the size nor the scope of this book permits. Here we can but express a pious opinion:—that they are individuals, that here once more the tendency towards the formation of *closed systems* has manifested itself, though again in very varying degrees, so that some of the systems show but a glimmer of individuality, others begin to let it shine more strongly through. That their individuality is no mere phantasm I think we must own when we find men like Dicey and Maitland (12, p. 304) admitting that the cold eye of the law, for centuries resolutely turned away, is at last being forced to see and to recognize the real existence, as single beings that are neither aggregates nor trusts, of *Corporate Personalities*.

This being so, it yet remains true that the state or society at large is still a very low type of individual: the wastage and friction of its working are only too prominently before our eyes. With the examples of what life has accomplished in producing our own bodies, we can never despair. But we must not be too far tempted by biological analogies: the main problem is the same, but the details all are new. The individuals to be fused into a higher whole are separate organisms with conscious, reasoning minds—personalities; and the solution will never be found in the almost total subordination of the parts to the whole, as of the cells in our own bodies or the sweated labourer in our present societies, but in a harmony and a prevention of waste, which will both heighten the individuality of the whole and give the fullest scope to the personalities of all its members.

THE RELATION OF INDIVIDUALITY TO MATTER; CONCLUSION

"Shall man into the mystery of breath,
From his quick-beating pulse a pathway spy?
Or learn the secret of the shrouded death,
By lifting up the lid of a white eye?"
MEREDITH.

A very striking experiment can be made on many of those free-living flatworms, the Planaria. If they are cut in two longitudinally, the halves will regenerate into perfect wholes, and this whether they are fed or not. If not fed they present us with a strange spectacle (Fig. 15). Without food, they cannot of course rebuild their missing block of buildings as we should, with new bricks: indeed, as energy has to be expended in the construction, some of the existing materials must be sacrificed as energy-producers, so that by the time the bit of worm-protoplasm has turned itself into a worm, it has actually decreased in bulk (Fig. 15). The half-worm has never ceased to exist as a half, but has somehow managed to become an ever smaller half while remodelling itself continually and at the same time handing over material for the building of what is missing. Finally the other half is completed—a whole worm has been made; up till now the old half had been decreasing rapidly in size, the new increasing almost as fast. From the

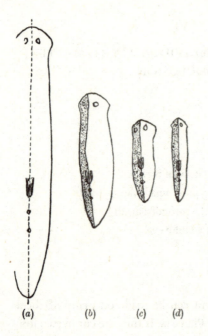

(a)   (b)   (c)   (d)

Figure 15
Planaria lugubris. Four stages in the regeneration of a whole from a
longitudinal half. The dotted line in (*a*) marks the line of the cut. The
stippled areas represent regenerated tissue. The figures are all drawn to
scale. (After Morgan.) (Slightly magnified.)

time that a whole is formed, both halves behave alike, decreasing slowly together as a result of starvation.

     This and many other similar facts tend to show that the
relation of form, and so (since specific form or structure is only
the visible machinery of a specific working) of individuality
in living things to its physical basis of matter, is primitively a
simple one, though one that is at variance with all our preconceived ideas. It seems to be this: any separate mass of one kind
of protoplasm will be able to, or rather must, make itself into
an individual with the form characteristic of the species. The

only provisions are that it is neither too large nor too small within certain defined limits and, of course, that the external conditions are favourable.

Facts suggesting this we have seen in Clavellina, in Stentor, and in Sycon:[1] indeed, as I have said, it really seems to be an original attribute of life, only more wonderful and startling than ordinary embryonic development because it is no regular part of the cycle of the species. Through its help the animal can extricate itself from positions in which it has never before been.

But, like most other primitive attributes of life, it has undergone considerable restrictions in the course of evolution. Animals, like men, cannot have their cakes and eat them. Three main factors have led to a restriction of this power of regeneration. The first is the formation of different substances for the performances of different functions, and their subsequent segregation into different regions of the body. These substances may get so specialized, so different from each other and from their common ancestor, that one cannot produce the other, and the presence of both is necessary in a mass of substance which is to give rise to a whole individual. In Stentor, for instance, although both nucleus and cytoplasm alike are living Stentor-protoplasm, yet a bit of one without the other will not regenerate. Here the two substances have been segregated by internal differentiation within the cell. Something similar occurs in Sycon, where the collar-cells by themselves cannot regenerate the other forms of tissue necessary to make a complete sponge; here aggregate differentiation has been at work, and whole cells and tissues are affected instead of parts of cells.

The second narrowing factor is harder to precise; but though we do not know its exact nature, we can often see it at work. There are many animals, such as man himself, where regeneration is almost non-existent although in any given case all the necessary substances and kinds of tissue would appear to be present. Here the failure to regenerate seems to stand in some general relation with the degree of specialization of the tissues; most animals can regenerate more completely when young or embryonic than when they are grown up.

The third factor is more obvious: certain bits of organic machinery are of such a nature that it is physically impossible for the animal to live at all if they are seriously tampered with. It is just because our blood-circulation is so swift and efficient and our nervous system so splendidly centralized that damage to heart or brain means almost instant death to us while a brainless frog will live for long, and a heartless part of a worm not only live but regenerate. Thus here again sacrifice is at the root of progress, and only by surrendering its powers of regeneration and reconstitution has life been able to achieve high individualities with the materials allotted her.

But this original property of living matter is important to us in one way. We begin to realize what an influence the cor-relation of parts can exert—how one part can affect others by its mere presence or absence. In Stentor, each bit that if sep-arated from the rest would grow into a perfect little whole, remains as a part as long as it is connected with the other parts. If it forms a part, it is because of its relation with other parts; if it forms a whole, it is because it is freed from that relation. Whatever it does, in fact, is due to the tendency of any sepa-rate mass of Stentor-protoplasm to form a whole Stentor.

Exactly similar is the behaviour of the blastomeres or sep-arate cells of the segmenting egg (pp. 52–53) only here the subordination is in one way more startling, for each of them is a single cell and represents historically a whole individual.

Similarly in all animals where small fragments can reconstitute miniature wholes, the fate of any particular cell in a fragment is determined very largely by its position in the fragment, and would be different if the fragment were of a different size or shape.

This "tendency towards wholeness" thus manifests itself across cell-boundaries as easily as through the more continuous substance of a single cell. More than this, it often seems to disregard them altogether. Many facts of embryology, as when form appears first and cells only later, lead us inevitably to a standpoint resembling that of Whitman (19), when he says of normal development:—"the plastic forces heed no cell-boundaries, but mould the germ-mass regardless of the way it is cut up into cells." Such considerations have led him and several others to throw up the cell-theory altogether, saying that the cells of a metazoan are not homologous with free-living protozoan individuals, but are merely convenient bricks, so to speak, or centres of local government, produced by the forces of life after the form of the creature had been established. But such a conclusion cannot be justified. We must carefully distinguish between what exists to-day, whether in adult body or developing embryo of a metazoan, and what we believe to have happened in the past.

Volvox and Haplozoon, whose cells we can with no shadow of doubt affirm to be homologous with free-living Protozoa, show that it is possible for a higher individual to be evolved from a collection of lower ones. If we refuse to the Metazoa an ancestor formed thus by aggregate differentiation, we are landed in far more and far worse difficulties than any we escape from. Whitman is right in drawing attention to the remarkable fact that the so-called Kupffer's vesicle of embryonic Teleost fish is non-cellular, a mere thin sheet of protoplasm which is not even nucleated, whereas it is certainly homologous with a structure of other vertebrates which is composed of very definite cells, but to reject the cell-theory altogether on this account is not warranted. Rather should we in such facts see examples of the extreme lengths to which the degradation of the individuality of the parts can go—a degradation which we found to be everywhere (except in man's societies) a necessary accompaniment of the formation of a higher individual from an aggregate. Here the cells have become degraded to the level of mere bricks, with even less share in determining the form of the whole than real bricks have in determining the form of a house. But how different is the structure of our Sponge or of Volvox—and they deserve equal consideration with the fish. It is better to believe in the historical individuality of the cells and to wonder at the idea of the whole's form that can thus penetrate the substance and absorb the individualities of its parts, robbing them of all their ancestral freedom, as the universal mind (some would believe) absorbs and loses in itself our souls at death. But here we have come down to the bed-rock questions of biology—the old problems of ordered growth and purposeful working, which are still shrouded in their dense cloud of ancient mystery.

Yet though, like enquirers who try to push far after knowledge in any direction, we are at length brought face to face with the unknown and perhaps unknowable, we have made some solid progress. Without discovering the origin or the inner being of individuality, we have been able to see it made objective in the various streams and masses of protoplasm which we call animals and plants and to trace an upward progress in its course, at the same time getting light on many related problems of biology. We have seen the totality of living things as a continuous slowly-advancing sheet of protoplasm, out of which nature has been ceaselessly trying to carve systems complete and harmonious in themselves, isolable from all other things, and independent. But she has never been completely successful: the systems are never quite cut off, for each must take its origin in one or more pieces of a previous system; they are never completely harmonious, as Metschnikoff's long list of the "disharmonies" in man will show; and they are never completely independent. These very incompletenesses, due to the limitations of the material stuff with which life has to work have proved the foundations of fresh advance. It is just because every system is bound to be in some degree dependent, that a number of systems can adjust their various ways of dependence to each other, till a condition of minimum wasted and maximum interdependence is gradually set up, and a new system, better equipped than any and all of the earlier ones, is made.

These systems are individuals, and it thus comes about that individuals exist in grade upon grade, any one in any grade being able to combine with others like itself or with others unlike itself to form the beginnings of a new system, a new individual. Moreover, within each grade there may exist individuals of every degree of perfection. At the bottom, a Gonium-colony is but a possibility of an individual; the individual formed by the inter-relation in food-matters of plants and animals is so vague as scarce to deserve the name. At the top, Man astounds by his harmonies, his purposeful completeness, and power over nature; but none are perfect. Thus we must not expect any hard-and fast rule; there are many grades, many degrees, and many kinds of individuality, and each individual must be judged on its merits, as something really new.

Finally we have learnt to appreciate the historical point of view, and through it to be brought to admire the seemingly

infinite changeableness of life. On the one hand we have seen
many structures and many habits of animals that can only be
made fully intelligible through their history. Each new species
must go through its period of storm and stress while striving
to come into harmony with its environment;

> "And 'mid this tumult Kubla heard from far Ancestral
> voices!"

—the forms and patterns of its forefathers rise up and will
not be denied, forcing themselves into the altered mould, and
thereby often taking on new and unfamiliar shapes.

The ancestral plan may persist in spite of present useless-
ness, like the elaborate arrangement of the lines of hair on the
body and limbs of man; or it may take on some new use, like
our Eustachian tube, in fish-like ancestors a gill-slit. It is by
this incorporation of the old in the new that we can trace such
adventurous histories as that of the cell-individual.

But this persistence is not absolute: with necessity and long
lapse of time life seems able to cast away every vestige of the
old forms, as when gills are replaced by lungs in air-breathing
vertebrates, or when a metazoan structure, once cellular,
builds itself without cells.

All roads lead to Rome: and even animal individuality
throws a ray on human problems. The ideals of active harmony and mutual aid as the best means to power and progress;
the hope that springs from life's power of transforming the
old or of casting it from her in favour of new; and the spur
to effort in the knowledge that she does nothing lightly or
without long struggle: these cannot but help to support and
direct those men upon whom devolves the task of moulding
and inspiring that unwieldiest individual—formless and blind
to-day, but huge with possibility—the State.

LITERATURE CITED

(1) BERGSON, H. "Creative Evolution" (Translated). London. Macmillan. 1911.

(2) CALKINS, G. N. "The Protozoan Life Cycle." Biol. Bulletin XI. 1906, p. 229.

(3) DARWIN, C. "Journal of Researches." London. J. Murray. 1888.

(4) ——— "The Variation of Animals and Plants under Domestication." London. J. Murray. 1875.

(5) DOBELL, C. C. "The Principles of Protistology." Archiv f. Protistenkunde XXIII. 1911, p. 269.

(6) DOGIEL, V. (Catenata.) Zeitschr. f. Wiss. Zool. LXXXIX. 1908, p. 417 and XICV. 1910, p. 400.

(7) ENRIQUES, P. (Conjugation, &c., in Infusoria.) Arch. f. Protistenkunde XII. 1908, p. 213.

(8) HUXLEY, T. H. "Collected Scientific Memoirs." London. Macmillan.

  (a) "Upon Animal Individuality." Vol. I. p. 146.

  (b) "On the Agamic Reproduction and Morphology of Aphis." Vol. II. p. 26.

(9) HUXLEY, J. S. (Regeneration in Sycon.) Phil. Trans. Roy. Soc. (B), vol. CCII. 1911, p. 165.

(10) KEEBLE, F. "Plant-Animals." Camb. Univ. Press. 1911.

(11) LE DANTEC, F. "Théorie nouvelle de la vie." Paris. F. Alcan. 1908.

(12) MAITLAND. Collected Papers. Vol. 3, pp. 285 and 302. Camb. Univ. Press. 1911.

(13) MORGAN, T. H. "Regeneration." New York. Macmillan. 1901.

(14) NEWMAN AND PATTERSON. (Armadillo Quadrupletes.) Biol. Bulletin XVII. 1909, p. 181.

(15) PERRIER, E. "Les Colonies Animales." Paris. 1881.

(16) ROUX, W. "Der Kampf der Theile im Organismus." Leipzig. 1881.

(17) WEISMANN, A. "The Evolution Theory" (Translated). London. Arnold. 1904.

(18) WHEELER, W. M. "The Ant-Colony as an Organism." Journ. Morphology, 1911, p. 307.

(19) WHITMAN, C. O. "The Inadequacy of the Cell-Theory." Biol. Lectures, Woods Hole, vol. II. 1893, p. 105.

(20) WOODRUFF, L. L. (Life-Cycle of Paramaecium.) Biol. Bulletin XVII. 1909, p. 287.

APPENDIX A

TABLE TO SHOW THE FIRST THREE GRADES OF INDIVIDUALITY; AND TO INDICATE THE DIFFERENCE BETWEEN ACTUAL AND HISTORICAL INDIVIDUALITY*

	(a) Functioning as wholes: *Actual Individuals.*	(b) Functioning as parts, but descended from ancestors that functioned as wholes; thus, though in point of fact not actual individuals, they are morphologically and historically equivalent to them, and may be called *Historical Individuals.*
(A_1) Individuals of the First Grade.	One of the hypothetical non-nucleated ancestral cells (p. 129n6).	A tissue-cell of Hydra or Man.
	A Protozoan.	A cell of Volvox or Haplozoon.
	A fertilized ovum.	A green cell in Convoluta.
(A_2) Compound wholes made up of first-grade individuals; without division of labour.	Gonium (p. 77).	

* Huxley's Appendix A is reprinted here in its original form. This first (a) column of the table accurately reflects the logic presented in the book, but the entries in the (b) column are puzzling because their positioning appears to contradict claims made in the body of the text. It is worth noting that this discrepancy disappears if the "Individuals of the First Grade" entries in the (b) column are moved down to "Full Individuals of the Second Grade" in the same column. The entries originally placed in this position would then be moved down to column (b) of "Full Individuals of the Third Grade." When these changes are made, the table becomes fully compatible with the analyses offered throughout the main text of the book. It is therefore tempting to speculate that typographical errors were introduced into Appendix A by the original publisher in 1912, but this hypothesis has yet to be validated through examination of the relevant archival material.

$(A_3)=(B_0)$ Ditto, but with division of labour (rudimentary second-grade individuals).	Volvox (p. 79). Haplozoon (p. 81).	
(B_1) Full individuals of the Second Grade.	(*a*) Clathrina (p. 68). Hydra (p. 30). Man, regarded singly. Pronuba, in certain respects (p. 97).	(*b*) A single polyp of Bougainvillea (p. 29), or of a Siphonophoran (p. 89). Man, regarded as a Unit of Society. Pronuba, in certain re-respects (p. 102). Convoluta, considered apart from its green cells (p. 96). A single Ant (p. 108).
(B_2) Compound wholes made up of second-grade individuals; without division of labour.	Many Sponges. Many Corals. Some Polyzoa (p. 131n1).	
$(B_3)=(C_0)$ Ditto, but with division of labour (rudimentary third-grade individuals).	Hydroid colonies such as Bougainvillea (p. 29). Polyzoa with avicularia (p. 131n1).	
(C_1) Full individuals of the Third Grade.	Siphonophora (p. 89). An Ant Community (pp. 9, 108). Human Society (pp. 108–109). Yucca-plant plus Pronuba (in certain respects) (p. 97). A Lichen (p. 93). Convoluta plus its green cells (p. 96).	

ON THE DIFFERENCES BETWEEN THE CELLS OF THE HIGHER PLANTS AND THE HIGHER ANIMALS

It is probable that in certain points the cells of the higher animals and the higher plants are not strictly homologous with each other.

Botanists distinguish three main types of elementary structure among plants, their differences arising out of differences in the method of cell-division practised (Fig. 16). In the first type (Coccoid), the entire cell, with its cell-wall, divides into two similar and quite separate halves. This is practised, e.g., by unicellular Algae. In the second type (Filamentous), the cell-body (cytoplasm and nucleus) divides as before, but the cell-wall does not divide; instead, an entirely new party-wall is laid down between the two cell-bodies, and in this partition small apertures are left, through which the two cell-bodies enjoy protoplasmic communication. This type of organization is found in all the higher green plants. In the third type (Coenocytic) the nucleus alone divides, and the final result is a *coenocyte*—a single overgrown cell with a single cell-wall and many nuclei. This plan has been adopted by the Siphoneae (pp. 67–68).

It is obvious that the first method is the most primitive and will be most generally practised by unicellular organisms; but whereas it has been abandoned by the higher plants, it seems to have been retained by the higher animals. Almost the only difference between the division of a protozoan and a metazoan cell lies in the fact that the two daughter-cells separate in the

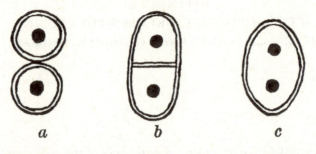

Figure 16
Diagram to show the three main types of elementary structure found
in plants. (*a*) coccoid, (*b*) filamentous, (*c*) coenocytic. In each case is
shown the sum of the changes following upon binary division of the
nucleus of a single cell.

one case, cohere in the other. The essential separateness of the
cohering cells is well seen in the collar-cells of simple Calcareous Sponges like Clathrina; here indeed there is even no continuity of coherence during normal life (pp. 70–71).

Similar if less strikingly separate cells can be seen in many
other groups of multicellular animals, and there can be very
little doubt that the first method of division was employed by
the common ancestor of all Metazoa;[1] true party-walls like
those of filamentous plants do not exist in animals, and animal syncytia (tissues formed by the coenocytic method) are
undoubtedly secondary.

We must now try and see what these facts mean. In the filamentous type the units are still homologous, as units, with the
original units we called cells (p. 43); but they have sacrificed
a considerable amount of independence. The whole mode of
division by which they arise is an obvious adaptation to a state
of existence where each is to be part of a continuous whole.

In Metazoa the separation of the cells is as a rule total, and if protoplasmic continuity exists, it appears to be secondarily produced. As regards their mode of cell-division, therefore, the Metazoa are more primitive than the Metaphyta; yet in spite of—or perhaps because of—this very separateness of their units, there has been a much greater division of labour between different kinds of units in animals than in plants.

To sum up: the cells of the higher plants and of the higher animals are both true cells—they are both broadly homologous with the original units of living matter. But the mode of cell-division in the two groups, in so far as it concerns the separation of the cells and the formation of the boundary between them, is *not* homologous.

CHAPTER I

1. There may appear to be a vicious circle in the use of the word individual before we know its definition; in reality there is not. The word individual has not been manufactured to label a theoretical concept, but to denote something existing. It was originally applied to human beings, and a special word had to be used for them because it was felt that they differed in certain important ways from mere *things*. Certain other objects (all of them organic, but together making only a portion of the whole organic world) are immediately recognized as possessing similar attributes, and it is obvious that they too must be Individuals, although equally obvious that we have only used, without defining, the category "Individual."

2. See pp. 65–68 for some further treatment of the value of size.

3. That is, of course, supposing the external world and the properties of matter allowed it to exist at all when in such small masses: e.g. Lillie has proved that there is a minimum size (determined no doubt chiefly by surface-tension) below which pieces of Stentor (a ciliated Infusorium) cannot regenerate. See pp. 36–37.

4. For a case of heterogeneous physical *structures* which cannot exist simultaneously, see pp. 82–83, 86. There the structures must alternate with each other in cyclical change; here, the memory obviates the necessity for that. Though two states of consciousness cannot actually co-exist at one moment of time, for all practical purposes memory permits it, as when we say that a man can attend to his profession and write a book upon some other subject "both at once," or as when a chess-player plays a dozen games "simultaneously."

5. For a fuller treatment of both these conceptions, see an article on "The Meaning of Death" in the *Cornhill Magazine* for April 1911.

6. As examples will serve, the hollow trunks of aged trees, the brittleness of old bones, and the decay of teeth.

7. It is to be noted that no actual impossibility stands in the way of the individual's continuance, but only great practical difficulties. Bergson somewhere makes the illuminating remark that the whole of Evolution might have realized itself in a single individual. This, with our knowledge of the potential immortality of many kinds of functioning protoplasm (see Metschnikoff on the age of trees, their propagation by cuttings, etc.) on the one hand, and of the facts of embryology, more especially the striking changes that take place at metamorphosis, on the other, we shall not readily be prepared to deny; but Life, gifted with reproductive powers has found it come cheaper and easier to choose Death for each single individual and think rather of the persistence of the race than to expend ever-increasing energy on patching up the defects that are bound to appear in the individual with age. (See *Conhill*, 1911, loc. cit.)

8. It would be more accurate to say *at least one* individual: often two or more distinct and unlike individuals are employed in each cycle of working; see p. 16.

CHAPTER II

1. Those which do not serve this end (with the exception of some which appear to be "accidental" by-products due to the interaction of the purposeful factors) are of course destined to help in reproduction.

2. Or at least without adding more than a very small amount. Normal growth may go on, but the re-modelling goes on still faster. That growth and reorganization are not necessarily connected is shown by the strange facts narrated on pp. 111–112.

3. The animal kingdom is divided into the two primary sub-kingdoms Protozoa or single-celled animals, and Metazoa or many-celled animals.

4. *Almost* universal, for it will be seen later that mental powers have made possible such organisms as an ant-colony, which is not a solid whole, single and defined in space; and growth and mobility may be in abeyance for long periods, though always present in some stage of an organism's life.

5. Though, as in commerce, one organism's manufactured article is another's raw material; take as an example the quadruple chain of nitrogen-fixing bacterium, clover, ox, and man.

6. Some biologists wish to restrict the term cell to protoplasmic units with a formed nucleus. The nucleus, however, has certainly arisen by internal differentiation (pp. 45–46), so that the lowest non-nucleated moneron, the complex protozoan, and the specialized metazoan tissue-cell are all *homologous*, and some word is required which will cover them all. Whether a mass of protoplasm is nucleated or not is of importance, but it is of still more importance to know whether it has arisen by a series of divisions from a primary unit of life, and so whether it is itself a primary unit of life, and so whether it is itself a primary unit. Such units we shall here call cells.

7. It will be objected that the change in surface-tension permitting binary fission is adaptive or purposeful. This is quite true, but the adaptation is concerned only with the race; it is the first step towards a species-individuality. The cells within the species, however, remain unaffected.

8. There is in many protozoa a form of circulation known as cyclosis, in which the whole inner part of the cell is constantly revolving. This certainly performs the same general functions for the organism as does a blood-system, but to have the *whole* of one's inside always in motion would render difficult the development of other systems; thus a huge single cell with cyclosis would have overcome the difficulty of metabolism, but would be at a disadvantage in other ways when compared with a multicellular organism of the same size.

9. In the vegetable kingdom, things are somewhat different, and the largest plants, the great forest trees, are individuals of a grade higher again than the second.

CHAPTER III

1. It is well known that there are two kinds of twins: *identical twins*, always of the same sex and almost indistinguishable from each other, and *ordinary twins*, which may be of opposite sexes, do not resemble each other more closely than brothers of different ages, and like them arise from the fertilization of two separate ova by two separate spermatozoa.

2. It is an interesting fact that the four twins fall naturally into two pairs, the resemblance between the members of which is still more close than that between the four taken together. This taken together with the fact that the members of the pairs are always adjacent seems to show that the fertilized egg divided into two halves, A and B, which did not remain united. Then A divided into $a1$ and $a2$, B into $b1$ and $b2$, and these again parted company. These four cells gave rise to four separate embryos, $a1$ and $a2$ forming one pair, $b1$ and $b2$ the other. Thus one pair is descended from A, the other from B, and the closer resemblance of the members of a pair is explained by closer blood-relationship.

3. The examples actually used by him are the Salpae and the Aphides.

4. In one species of Pilidium (*P. recurvatum* Fewkes), however, the young worm does actually absorb the remains of the larva. It is interesting to note that a precisely similar series can be traced in the metamorphosis of echinoderms. In the sea-cucumbers the process is almost entirely one of remodelling, in sea-urchins and most starfish the young imago is formed apparently as a "bud" and the rest of the larva is absorbed later, while in some starfish (*vide* J. Müller) the larva and the imago part company.

5. As a matter of fact there are animals, such as the sea-urchins, where death results from division of the body and yet is certainly not caused by any dislocation of nervous centres, for the sea-urchins have a very feeble and very decentralized nervous system.

CHAPTER IV

1. I am aware that botanists distinguish between *cells*, which have one nucleus, and *coenocytes*, or masses of protoplasm with many

nuclei, such as are found in Caulerpa and other Siphoneae. However, I am using the word cell in a wide sense, a sense dictated by the historical or evolutionary point of view, to denote a discrete mass of protoplasm isolated by natural causes, and if this definition be allowed, then Caulerpa is simply a single cell which has found out the way to become large. The number of nuclei in a cell is often quite unimportant: in the Protozoa one form may have a single nucleus, while a close relation has several.

2. But not unique—e.g. in some colonial Ascidians, the germ-cells of the bud are formed from blood-corpuscles of the parent.

3. This has been done on various sponges, including Sycon, a not very distant relation of Clathrina: see Huxley (9)

4. Some species of Gonium, such as that represented in Fig. 7, are even simpler, being formed of but four cells.

5. Though these connections have not been described for other members of the family, it is possible that they have been overlooked.

6. To mention two examples, there is the strobila with its ephyrae, and the Syllids producing their special (epitokous) male and female forms by division.

CHAPTER V

1. It is interesting to note that in the Polyzoa, another group of colonial animals there has been a different kind of division of labour. The ordinary animals both feed and reproduce the colony, and defence is undertaken by much modified persons called Avicularia (from their resemblance to birds' heads with snapping beaks). Here the differentiation is between most of the somatic and all the germinal functions on one side, and a single somatic function on the other. In some forms there are no Avicularia, the colonies then consisting of only one kind of person.

2. A metaphorical effort, as when a carnivorous species acquires new powers of speed to run down its prey, or an actual effort, as when the members of that species make use of those powers.

3. See Dobell (5) for facts and references.

CHAPTER VI

1. pp. 35–37 and 71, respectively.

APPENDIX B

1. It is more than probable that Sponges have an ancestry quite separate from the rest of the Metazoa: if so, then the common ancestor of Sponges employed, though quite independently, the same method as the ancestor of the Metazoa proper.

INDEX

Note: Page numbers in italics refer to figures.

spatial, 18
of a species, 15–16, 18, 62
temporary, 96–97
tendencies and progress of,
    21, 88
various definitions of, 25, 51,
    62–64
Internal differentiation, 46,
    103–104, 105–106

Jelly-fish
    artificial production of twins
        and quadruplets in, 53
    lack of complexity, 5
    reproductive in function in
        Hydroids, 89–91
Jerboa, thigh-muscles, 66–67

Kangaroo, size of, 66–67
Keeble, 96
Keller, Helen, 100
Kite, used to bring lightning to
    earth, 49
Kubla, 118
Kupffer's vesicle, 115

Le Dantec, definition of
    individual, 62–63
Lichens, compound species, 93,
    103, 122
Limbo, 26
Liriope, a jelly-fish, 53
Liver Fluke, 16–17, *17*, 62

Maitland, 109
Malaria, 4
Man, 114, 117, 121, 122

communities of, 49, 85,
    108–109
great independence of, 5
and individuality, 25–28, 53
the tool-maker, 9–10
Materialism, errors of, 64
Medusae, 89–91
"Merrimac" and "Monitor,"
    first armoured ships, 87
Metamorphosis, 55–60, 128n7
    reason of, 58–60
Metazoa
    compound individuals, 28–
        29, 34
    and protozoa, 34
Metschnikoff
    and death, 128n7
    and disharmony, 117
Microscope, 4
Milton, and life before birth,
    26–27
Minoan dancers at bull-fights,
    14
Monsters, double, 52
Mutations, and individuality,
    61

Nectarine, produced as bud-
    sport from peach, 62
Nelson, 87
Nemertine worms,
    metamorphosis in, 55–57, 60
Nero, 43
Nerve-cell, 104
Nervous system, 48
    supposed basis of
        individuality, 63